Children's Mathematical Thinking

A DEVELOPMENTAL FRAMEWORK
FOR
PRESCHOOL, PRIMARY, AND SPECIAL EDUCATION TEACHERS

Children's Mathematical Thinking

A DEVELOPMENTAL FRAMEWORK FOR PRESCHOOL, PRIMARY, AND SPECIAL EDUCATION TEACHERS

Arthur J. Baroody

CARL A. RUDISILL LIBRARY
LENOIR-RHYNE COLLEGE

Teachers College, Columbia University
New York and London

Grateful acknowledgment is given for permission to use previously published material from the following sources:

Baroody, A. J. (1984). Children's difficulties in subtraction: Some causes and cures. *Arithmetic Teacher, 32*(3), 14–19.

Baroody, A. J. (1984). Children's difficulties in subtraction: Some causes and questions. *Journal for Research in Mathematics Education, 15*(3), 203–213.

Baroody, A. J. (1986). The value of informal approaches to mathematics instruction and remediation. *Arithmetic Teacher, 33*(5), 14–18.

Baroody, A. J. (in press). The development of computational procedures for single-digit addition. *Journal for Research in Mathematics Education.*

Baroody, A. J., & Ginsburg, H. P. (in press). The relationship between meaningful and mechanical knowledge of arithmetic. In J. Herbert (Ed.), *Conceptual and procedural knowledge: The case of mathematics.* Hillsdale, NJ: Lawrence Erlbaum Associates.

Baroody, A. J., & Ginsburg, H. P. (1983). The effects of instruction in children's concept of "equals." *The Elementary School Journal, 84,* 199–212.

Number: The Language of Science, Fourth Edition by Tobias Danzig. Copyright 1930, 1933, 1939, 1954 Macmillan Publishing Company, renewed 1958, 1961, 1967 by Anna G. Danzig. Reprinted by permission of the publisher.

Published by Teachers College Press, 1234 Amsterdam Avenue, New York, N.Y. 10027

Copyright © 1987 by Teachers College, Columbia University

All rights reserved. No part of this publication may be reproduced or transmitted in any form or by any means, electronic or mechanical, including photocopy, or any information storage and retrieval system, without permission from the publisher.

Library of Congress Cataloging-in-Publication Data

Baroody, Arthur J., 1947–
Children's mathematical thinking.

Bibliography: p.
Includes index.
1. Mathematics – Study and teaching (Elementary)
2. Learning. I. Title.
QA135.5.B2846 1987 372.7 86-32823

QA
135.5
.B2846
1987
148879
May 1990

ISBN 0-8077-2838-1
ISBN 0-8077-2837-3 (pbk.)

Manufactured in the United States of America

92 91 90 89 2 3 4 5 6

For my parents, Arthur and Martha Keeley Baroody,
and my wife, Sharon Coslick Baroody

Contents

Foreword

Art Baroody's new book, *Children's Mathematical Thinking*, is a splendid contribution. I take great pleasure in recommending it as an invaluable resource for all those concerned with the mathematics education of young children.

Baroody's book is unusual in its simultaneous concern with the psychology of children's mathematics and the teaching of school arithmetic. This concern involves three major components. The book presents the educator with a powerful and practical framework for understanding children's learning of elementary mathematics. It then uses this psychological framework to generate some general guidelines for teaching. In addition, the book provides specific activities, often in the form of games, designed to promote the learning and teaching of elementary mathematics. Indeed, Baroody accomplishes the rare feat of producing a blend of the psychological and educational that is both theoretically sophisticated and eminently practical.

Baroody's psychology is a "cognitive" one, drawing on many years of research on children's mathematical thinking. The bulk of the research, much of which was conducted by Baroody and his colleagues, involves the intensive study of learning and thinking in individual children. Many of the observations presented in the book involve Baroody's own children, observed closely over a period of several years in the friendly surroundings of the home. Other observations involve case studies of children with learning difficulties whom Baroody treated in a clinical setting. Research of this type results in a rich and interesting psychology, obviously and directly relevant to the concerns of practicing educators. And it results, too, in a psychology that is easy and pleasurable to read!

Baroody's psychology contains several major propositions of value to the mathematics educator. One is that even before entering school, most young children possess an "informal mathematics" of some power. They possess, for example, some simple notions of more and less, counting, and adding. A second proposition is that young children often use their informal knowledge as a vehicle for interpreting the formal arithmetic that is taught in school. Thus, they may see adding as counting—even if the curriculum presents another view, for example, an exposition of adding as the union of two sets. In a sense, children *construct* their mathematical knowledge. As a result, a third proposition is relevant: Learning difficulties can arise if there is a gap between formal instruction and informal knowledge or if informal knowledge has not been mastered. Fourth, understanding children's work in mathematics, then, requires more than considering their right and wrong answers; it requires insight into their informal mathematical thinking, which often takes surprising and unexpected forms. The book shows in rich and interesting detail how informal mathematics develops, how children construct a sensible (from their point of view) school arithmetic, and how mathematical thinking operates.

This psychology suggests many general principles of teaching. One of the most powerful is that the educator can gain a great deal from knowing, respecting, and building upon the child's informal knowledge of mathematics. Often children experiencing great difficulty in school arithmetic will be found to possess unsuspected strengths in the area of informal mathematics. Thus, a child may have trouble with column addition but may be skilled at informal mental addition. This knowledge argues against a diagnosis suggesting low intelligence or poor mathematical aptitude. The approach Baroody advocates often results in the discovery of hidden potential in children who were previously assumed to have little chance of successful learning. The general principles emerging from the psychology of children's mathematics suggest new conceptions of school learning generally and of learning disorders in particular, and valuable approaches to teaching, remediation, and testing.

Finally, Baroody's book offers specific learning and teaching activities that build on the psychology of children's mathematics to promote sound calculational procedures, understanding of mathematical principles, mastery of number facts, and methods of problem solving. These activities often take the form of enjoyable games, and they have proved successful with many of the children Baroody has studied.

The blend of psychological theory and educational practice that Baroody presents is extremely rare and represents a vital contribution to mathematics education. This book should be an essential element in the training of teachers.

HERBERT P. GINSBURG
Teachers College, Columbia University

Preface

Knowledge of basic mathematics is an indispensable tool in our society. Counting objects, reading and writing numerals, performing arithmetic calculations, and reasoning with numbers are aspects of many of even the simplest tasks that confront adults every day. In addition to its importance as a survival skill, basic mathematics is the foundation for the scientific and more advanced mathematical knowledge required for many jobs in our technologically advanced society. Because of its fundamental importance, mathematics is a principal objective of primary school and special education alike. Even many preschool programs have a mathematics component.

This book is based on the premise that psychology can be an invaluable tool to mathematics educators. To help build a solid foundation of mathematical knowledge, preschool, primary, and special education teachers need to know how children learn mathematics and why they do not learn mathematics. In the past, because theory was either overly simplistic or not relevant to school mathematics learning, psychology failed to provide educators with sound ideas and techniques for helping children learn mathematics. With the recent emergence of cognitive theory, psychology is now in a position to really help educators understand children's mathematical learning and learning difficulties. This new theory and research can provide a powerful and practical framework for the long-range and daily planning of mathematics educators.

This is not to say that psychology can now provide all the answers or cookbook recipes for every situation confronted by mathematics educators. As William James (1939), the famous psychologist, cautioned educators in a talk at Teachers College of Columbia University in 1898:

> I say moreover that you make a great, a very great mistake, if you think that psychology, being the science of the mind's laws, is something from which you can deduce definite programmes and schemes and methods of instruction for immediate schoolroom use. Psychology is a science, and teaching is an art; and sciences never generate arts directly out of them-

> selves. An intermediary inventive mind must make the application, by using its originality. (pp. 7–8)

By applying the latest psychological knowledge, however, educators can make more informed and artful plans.

Part I of the book outlines general implications that stem from cognitive theory. This is intended to provide a general framework for analyzing mathematics learning (Chapters 1 and 2), teaching practices (Chapters 3 and 4), and learning difficulties (Chapters 2, 3, 4, and particularly 5). Parts II and III delineate more specific implications and guidelines of cognitive theory and research. Part II (Chapters 6 to 8) focuses on informal mathematics: the counting, number, and arithmetic knowledge that children typically learn before or outside the formal school setting. Part III (Chapters 9 to 14) describes school-taught or formal mathematics. Parts II and III outline models of how particular skills and concepts develop. These models provide a specific framework for understanding mathematical learning and how learning problems develop, devising instruction, and remedying learning difficulties. Instructional activities are detailed throughout Parts II and III.

The general and specific recommendations and activities described throughout the book are offered as *guidelines* rather than dictums. These guidelines are intended as tools to help mathematics educators think about the complex, challenging, and interesting task that confronts them. Some recommendations and activities may not be appropriate for every child or every situation. Thus the guidelines need to be implemented thoughtfully. To paraphrase William James: A judicious and creative intermediary must make the application.

Acknowledgments

Many resources and people, only some of which are named below, were instrumental in the creation of this book. The book is based, in part, on work supported by the National Institute of Child Health and Human Development of the National Institutes of Health (Research Grant HD 16757) and National Science Foundation (MDR-8470191). I am indebted to the B.O.C.E.S. I and II (Monroe County), Brighton, Rochester City, Penfield, Penn Yan, and Pittsford school districts in New York for their cooperation. My research and the book manuscript were made possible by the tremendous support facilities of the University of Rochester.

I would like to thank my colleagues Barbara Allardice, Harriet Bebout, Jamie Campbell, Robert B. Davis, William Doll, Jr., Karen Fuson, Vicky Kouba, and Lloyd Wynroth, and Teachers College Press editors Audrey Kingstrom and Peter Sieger for their helpful comments on the book. I am especially grateful to my mentor, Herbert P. Ginsburg, for his direct and indirect contributions to this project. Thanks are due to my research assistants who have contributed in various ways to the book: Bill Boaz, Brian Burley, Fred Dawson, Laurie Donofrio, Margaret Hank, Beth Messner, Rubina Saigonal, and Marianne Young. Special thanks are due Cathy A. Mason. I am most thankful for the skilled and thoughtful word processors Margaret Davidson, Betty Drysdale, and Judy Gueli, who diligently translated sometimes jumbled handwritten copy into print. Last but not least, I would like to thank all the children I have worked with and learned from, including my own Alexis, Alison, and Arianne.

Part I

GENERAL FRAMEWORK

CHAPTER 1

Two Views of Learning

Why is it important for educators to understand how children learn mathematics? What are the consequences of not adequately considering how children learn mathematics? What is the nature of children's mathematical learning? How is this learning process analogous to a mathematician's endeavors?

THE NEED FOR A PSYCHOLOGICAL UNDERSTANDING OF LEARNING

Educators are called upon daily to make decisions concerning both general and specific issues in all phases of mathematics education: curriculum, instruction, testing, and remediation. Consider the examples below.

- A principal is told that a publisher's latest textbook series and curriculum materials provide new and effective methods for teaching "basic skills." In terms of mental development, what are the basic skills? Does the curriculum package introduce the skills in a meaningful manner and in a developmentally sound sequence?
- A kindergarten teacher assigned to a curriculum committee is asked to evaluate the existing kindergarten program and make recommendations for improvements. Does the current kindergarten curriculum either underestimate or overestimate the mathematical knowledge of children just beginning school? Does it spend enough time developing readiness skills and concepts and introduce mathematics in an interesting and understandable manner?
- A first-grade teacher notices that many pupils solve their arithmetic assignments by counting on their fingers. Will finger-counting procedures impede or facilitate their mathematical development? Should their finger counting be discouraged or encouraged?
- A special education teacher is revising individual education plans (IEPs)

and notices that Jenny, a mildly retarded child, has had the same mathematics objective for three years: master the basic facts to sums of 18. Why has Jenny not memorized her basic addition facts, even after so much practice? Should the teacher continue to drill the facts or use a different instructional approach?

- A committee is assigned the task of evaluating the kindergarten screening test and recommending changes. Does the existing screening test do a good job of identifying children at risk, gauging skills essential to success in school mathematics, and providing a profile of specific strengths and weaknesses? If not, what test items could be quickly administered to accomplish these goals?
- Fred, a third grader who has been classified as learning-disabled, has just arrived from another school district. He is sent to the school psychologist for further evaluation. The school psychologist notices that, according to the KeyMath Test, Fred is two years below grade level in calculational ability and numeration skills, such as reading numerals. What are Fred's specific computational difficulties? Does he have any strengths upon which remedial efforts can build? How can the school psychologist quickly and systematically obtain specific diagnostic information that will provide clear directions for remedial efforts? Is it likely that Fred's numeration and computational problems are due to a perceptual-motor difficulty and that a test of perceptual-motor ability will provide instructionally useful information?
- A curriculum supervisor notices that many fourth and fifth graders in the school district still do not really understand place value. For example, some do not realize that two-digit numerals such as 73 can be thought of as 7 tens and 3 ones, and many have trouble with arithmetic involving two-digit terms. What suggestions for remedial work can the supervisor suggest to the district teachers?
- Sue, a second grader, has been sent to her school's math lab because she is having a very difficult time with her subtraction assignments. The remedial-math teacher asks Sue to calculate the answer to 52 – 35. Sue writes down an answer of 23! Why has Sue not mastered the standard procedures for subtraction? Why does she make the kind of error that she does? How should the remedial-math teacher proceed with Sue?

To make effective decisions, these educators need to understand how children learn mathematics. Psychological knowledge can assist educators in judging the appropriateness of the methods, materials, and sequence of a curriculum. An understanding of the learning process can help teachers decide how to introduce a topic and foster mastery. Knowledge of psychology can provide guidance on what is important to examine and on how to evaluate progress. Knowledge of the child can aid educators in anticipating where and why children will run into difficulty and how to prevent or remedy learning difficulties.

Indeed, it is imperative that educational planning take into account the psychology of the child. Whether making decisions about general or specific issues of curriculum, instruction, testing, or remediation, it is essential to consider how children learn and think (cognitive factors) and what they need, feel, and value (affective factors). If we do not take adequate heed of how children learn and think, we run the risk of making early mathematics education unduly difficult and discouraging for children (Brainerd, 1973). When school mathematics is taught with little regard for cognitive factors, many children will learn and use mathematics mechanically or mindlessly, and some will develop learning difficulties. Moreover, if the nature and pace of instruction are not suited to the child, there are often terrible affective repercussions in the form of debilitating emotions.

Consider the cartoon in Figure 1.1. Peppermint Patty is in the midst of an assignment in formal mathematics (school-taught mathematics based largely on written symbols). Typical of children, she has her own ideas and methods (informal mathematics) based on counting: $7+3=\_$ "is easy because you just take the first number and then count the little pointy things on the '3' [7; 8, 9, 10], and you have the answer!" When confronted with problems with large numbers—a task for which she is not ready—her whole demeanor and approach change. She feels helpless ("No one can be expected to answer a problem with a 'twelve' in it!"). Though she knows it does not make sense, Peppermint Patty responds mechanically ("If a problem has really big numbers in it, the answer is always 'one million'!"). When formal instruction is not geared to their level, children tend to perceive mathematics as difficult, mysterious, and even threatening ("Math is like learning a foreign language, Marcie . . . no matter what you say, it's going to be wrong anyway!"). This contrasts vividly with the ease, confidence, and enthusiasm with which children use their informal mathematics ("'Nine plus three' . . . I take the nine and count the little pointy things on the three . . . ten, eleven, twelve . . . the answer is 'twelve' . . . Ha!!").

To make informed decisions about mathematics instruction, educators need a powerful psychological framework that can deliver an accurate explanation of learning. For example, a powerful theory provides a sound account of how and why children master the basic number combinations, such as $9+0=9$ and $5+5=10$. Because there are different theories of learning that have very different educational implications, it is essential that educators base their decisions on the most powerful theory available.

Though we may not consciously view it as a theory, everyone has a set of

FIGURE 1.1: Peppermint Patty's Thoughts on Mathematics

© 1980 United Feature Syndicate, Inc.

beliefs about how mathematics is learned. These beliefs have an influence on all aspects of teaching: They govern what is considered appropriate to include in a curriculum and when topics should be taught; they determine the importance an educator attributes to gauging readiness skills or exploiting children's curiosity and interests; and they affect how educators teach skills and concepts, evaluate progress, and remedy difficulties. In brief, whether conscious or not, beliefs about mathematics learning guide decision making and, in the end, influence our effectiveness as mathematics educators. Thus it is essential that an educator carefully examine his or her view of learning.

TWO THEORETICAL APPROACHES

Basically there are two general theories of learning, absorption theory and cognitive theory. Each reflects a different belief about the nature of knowledge, how knowledge is acquired, and what it means to be knowledgeable. Absorption theory suggests that knowledge is impressed upon the mind from without. Basically, knowledge is viewed as a collection of facts. Facts are learned by means of memorization. In effect, learning is a process of internalizing or copying information. A knowledgeable person is someone with a good store of memorized information and ready recall. Cognitive theory argues that meaningful knowledge cannot be imposed from without but must be worked out from within. Genuine knowledge entails insight or understanding. Meaningful learning is a different process from learning by rote memorization (Katona, 1940/1967). Learning by insight or understanding is effectively a problem-solving process: noting and then puzzling over clues, rearranging the available evidence, and finally seeing a problem in a new light. A knowledgeable person is someone with insight and the means to solve new problems. To examine each theory in more detail, we turn now to how each accounts for an elementary aspect of mathematical learning: mastery of the basic number combination (e.g., the single-digit addition facts $0 + 0 = 0$ to $9 + 9 = 18$).

Absorption Theory

ASSOCIATIVE LEARNING. According to absorption theory, mathematical knowledge is essentially a basket of facts and skills. At the most basic level, learning facts and skills involves forming associations. For example, mastering a basic addition combination requires learning that a number pair is associated with a particular sum (e.g., that 7 and 3 are associated with 10). Automatic and accurate production of a basic number combination is simply a well-ingrained habit of associating a particular response with a particular stimulus. For instance, upon seeing or hearing the stimulus 7 + 3, the child looks up the associated sum in long-term memory and responds, "10." In brief, absorption theory assumes that mathematical knowledge is a collection of facts and habits made up of basic elements called associations.

PASSIVE, RECEPTIVE LEARNING. In this view, learning involves copying facts and skills—an essentially passive process. Associations are impressed upon the mind largely through repetition. In plain terms, "practice makes perfect." For example, children cement the bond between 7 + 3 and 10 through repeated practice. With sufficient exposure, the fact 7 + 3 = 10 is firmly imprinted on the mind. Understanding is not deemed necessary for the formation of associations. A learner merely needs to be receptive and willing to practice. Put differently, learning is fundamentally a process of memorization.

ACCUMULATIVE LEARNING. According to absorption theory, knowledge growth involves building up a storehouse of facts and skills. Knowledge expands by the memorization of new associations. For instance, mastering the basic addition combinations involves accumulating 100 associations or facts. Furthermore, basic facts or habits can be linked together to form more complex facts or habits. Take mastering an addition algorithm (a step-by-step procedure) for two-digit addition without carrying (e.g., 46 + 23 = __). This skill entails stringing together six simple habits into a habit sequence: (a) start with the right-hand column, (b) retrieve the sum for the two terms in this column (6+3=9), (c) record the sum beneath these terms, (d) move to the left-hand column, (e) recall the sum of these two terms (4 + 2 = 6), and (f) record the sum beneath these left-hand digits. In other words, knowledge expansion is basically an accumulative process: an increase in the number of associations stored.

EFFICIENT AND UNIFORM LEARNING. Absorption theory assumes that children are simply uninformed and can readily be informed. Because associative learning is a straightforward copying process, it should be accomplished quickly and faithfully. Moreover, because all but the brightest and dullest have similar memorization abilities, learning should proceed at a relatively even rate. For example, because it appears to be nothing more than the memorization of some facts, children are expected to achieve adultlike proficiency with the basic number combinations in short order. The New York State Education Department, for instance, suggests in its curriculum guideline that children master sums and differences to 18 by the end of first grade. As long as facts and skills are presented clearly and practiced sufficiently, all but atypical children should proceed efficiently and uniformly toward mastery. In effect, rote memorization of knowledge should effectively be accomplished en masse in short order.

EXTERNAL CONTROL. Absorption theory assumes that learning must be controlled from without. To produce a correct association or a true copy, a teacher must shape a pupil's response by using rewards and punishments. Without the promise of reward or threat of punishment, children become inert. By controlling the rewards and punishment they receive, an educator can overcome children's natural reluctance to learn. For example, mastery of a "combination family," such as the times three (1×3, 2×3, 3×3, 4×3, and so forth), might be rewarded by placing a star after a child's name on a "times chart." Presumably this also works as an incentive for those who do not get the star. In essence, the motivation for and control over learning are external to the child.

Cognitive Theory

RELATIONSHIPS: A KEY BASIS OF LEARNING. In contrast, cognitive theory claims that knowledge is not simply a basket of facts (Anderson, 1984). The essence of knowledge is structure: elements of information connected by relationships to form an organized and meaningful whole. Thus, the essence of knowledge acquisition is learning general relationships. Try this exercise. Give yourself 60 seconds to learn the following "body of knowledge," an 11-digit number: 25811141720. Now cover up the 11-digit number. Can you recall the number? How did you set about learning the body of knowledge? Adults frequently have no problem with this exercise. Typically, they treat the exercise as a memorization task. However, few people try to memorize each term (11 facts). Instead, to make memorizing easier, they create chunks of information by grouping terms. Some people memorize many small chunks (e.g., 25-81-11-41-72-0 or 25-81-11-41-720). Some people memorize a few large chunks of information (e.g., 258-111-417-20 or 2581-1141-720). In any case, most people treat the body of knowledge as meaningless pieces of information (unconnected facts) that must be rotely memorized.

A rote-memory approach can work as long as a body of knowledge remains small. However, if our example of a body of knowledge is expanded—say, to 23 digits (25811141720232629323538)—the task becomes rather difficult, even with chunking. As a body of knowledge becomes larger, it will sooner or later outstrip our capacity for rote memorization. However, whether it extends for 11, 23, or 240 digits, we can easily remember this body of knowledge if we can discern its structure (Katona, 1940/1967). Our example is what is called an arithmetic progression: It starts with 2 and each successive number is determined by adding 3. Hence the second term is 5; the third, 8; the fourth, 11; the fifth, 14; and so on. Once we discover a relationship, we have a powerful tool for remembering a body of knowledge despite its extent.

Cognitive theory points out that, typically, memory is not photographic. We usually do not make an exact copy of the external world and store every detail or fact. Instead, we tend to store relationships that summarize information about many particular cases. In this way, memory can store vast amounts of information efficiently and economically.

Consider an elementary body of knowledge: the basic "number facts." For the four arithmetic operations (addition, subtraction, multiplication, and division), a rote-memory approach would mean learning and storing just over 300 specific numerical associations or individual facts. This makes mastering the combinations unduly difficult. However, the basic number combinations are not simply a basket of isolated facts. Important mathematical relationships underlie the basic number combinations. For example, the 20 basic combinations that include zero as an addend (e.g., $2 + 0 = 2$, $9 + 0 = 9$, $0 + 3 = 3$, $0 + 7 = 7$) are all embodiments of the same underlying relationships: Whenever zero is an addend, the other addend remains unchanged ($N + 0 = N$ or $0 + N = N$ rule). Learning this general relationship can enable a child to respond quickly and accurately to the 20 basic zero problems as well as an infinite number of other zero problems (e.g., $1{,}000{,}001 + 0 = \_$, $0 + 728 = \_$). Such general relationships summarize many particular cases and provide a powerful basis for storing and "remembering" an otherwise huge body of information.

ACTIVE CONSTRUCTION OF KNOWLEDGE. Cognitive theory proposes that genuine learning is not simply a matter of absorbing and memorizing information imposed from without. Insight requires thought. Understanding is actively constructed from within by relating new information to what is already known or by noticing a relationship between previously known but isolated pieces of information.

Connecting new information to existing information—understanding new information in terms of extant knowledge—is called *assimilation*. For example, I worked with a kindergarten girl who was unfamiliar with the symbol for zero. Puzzled by the problem $6 + 0$, the girl finally responded, "60." After I explained that 0 was the number name for adding "nothing," she answered that $6 + 0$ must be 6. Later in the same session, without further help, she quickly and correctly responded to $3 + 0 = \_$. A week later, she continued to respond quickly and correctly to zero problems. Though she apparently had seen zeros in two-digit numerals such as 60, the girl did not really understand the formal symbol 0 and, hence, arithmetic sentences containing the foreign symbol. By relating the unfamiliar zero symbol to her previous knowledge that adding nothing to a set leaves the set unchanged, she quickly assimilated the new symbolic information. As a result of her insight, the girl could then understand and respond appropriately to similar but new expressions such as $3 + 0 = \_\_$.

New understanding can also occur by means of *integration*: connecting previously isolated bits of information. For instance, consider the child who knows that he has 5 fingers on each hand and 10 fingers altogether, yet persists in calculating the sum when given the symbolic problem $5 + 5 = \_\_$. For this child, the practical knowledge concerning his fingers is not connected to his formal knowledge of addition. If these two bits of isolated knowledge can be brought together, the child will be able to respond automatically with "10" whenever he sees or hears the symbolic problem $5 + 5 = \_\_$. In summary, the growth of meaningful knowledge, either by assimilating new information or by integrating existing information, involves active construction.

CHANGES IN THINKING PATTERNS. Cognitive theory suggests that knowledge acquisition entails more than just accumulating information. Genuine learning involves changing thought patterns. More specifically, making a connection can change the way knowledge is organized, thus changing *how* a child thinks about something. Put differently, insights can provide fresh and more powerful perspectives. Take the child who does not know the basic subtraction combinations and must use finger counting to calculate differences. Given the series of problems $2 - 1 = \_\_$, $4 - 2 = \_\_$, $6 - 3 = \_\_$, $8 - 4 = \_\_$, $10 - 5 = \_\_$, the child laboriously calculates each answer. Suddenly the child has an insight: The subtraction combinations are a mirror image of the well-known addition doubles ($1 + 1 = 2$, $2 + 2 = 4$, $3 + 3 = 6$, $4 + 4 = 8$, and $5 + 5 = 10$). There is a relationship between subtraction combinations and the familiar addition facts! Afterward, she views subtraction in a different light. Given a problem like $5 - 3 = \_\_$, the girl now thinks to herself: "Three plus what makes five? Oh, yeah, two." Her new perspective now enables her to solve subtraction combinations efficiently without laborious calculation. Mathematical development, then, entails qualitative changes in thinking as well as quantitative changes in the amount of information stored. Essential to the development of understanding are changes in thinking patterns.

LIMITS ON LEARNING. Cognitive theory cautions that because children do not simply absorb information, there are limits to their teachability. Quick, faithful, and uniform imitation of adult knowledge is not realistic. Because of the processes of assimilation and integration, most things worth knowing take a long time to learn (Duckworth, 1982). Children construct their understanding of mathematics slowly, insight by insight. For instance, children only slowly learn the mathematical relationships that permit them to master the basic number combinations.

Because knowledge is actively constructed, children's ideas and methods for solving problems may not correspond to those prescribed by instruction. For example, despite the emphasis in primary school on memorizing the basic addition and subtraction facts, children typically rely on their own invented procedures to do basic arithmetic because, initially, that is more meaningful (e.g., Brownell, 1935). Based on what they do know, children spontaneously invent finger-counting strategies and "thinking strategies" to figure out unknown combinations (e.g., Steinberg, 1985). For instance, children will use known facts to reason out unknown combinations. For $5 + 6$, a child might use the known "double" $5 + 5 = 10$. The child then reasons that six is one more than five, so $5 + 6$ must be one more than $5 + 5$ or 11. This double-plus-one thinking strategy can be used to reason out many hard-to-remember combinations, such as $3 + 4$ or $4 + 3$ ($3 + 3$ is 6 plus 1 more is 7) and $6 + 7$ or $7 + 6$ ($6 + 6$ is 12 plus 1 more is 13).

Because assimilation and integration involve making connections with existing knowledge, meaningful learning is necessarily dependent on what an individual already knows. What may be obvious to one child may be unfathomable to another. For example, one child may readily understand the rule: adding zero leaves a number unchanged. Because it fits into her pattern of thinking, the child remembers the rule and can use it to solve written problems like $5 + 0 = \_\_$. The same rule may not make sense to a less advanced and experienced child. Because the rule does not fit into her pattern of thinking, the child dismisses the information and fails to solve written zero problems. Understanding and meaningful learning, then, depend on individual readiness.

INTERNAL REGULATION. Cognitive theory suggests that learning can be its own reward. Children have a natural curiosity—a natural desire to make sense of the world. As their knowledge builds up, children spontaneously seek out increasingly difficult challenges. Much is often made of the short attention span of young children. Indeed most young children readily abandon tasks that they find uninteresting. Yet when working at problems that interest them, children will devote considerable time—hours, and even days—to working toward mastery. For instance, children are often unenthusiastic or even unresponsive to number-fact drill. In contrast, they are excited when they discover or are helped to discover thinking strategies. In effect, mastering the basic combinations becomes a "detective game"—a problem-solving exercise. Because they are willing to invest themselves in this meaningful and interesting task, mastery can be facilitated.

Evaluating the Theories with Respect to Mathematics

Which theory provides the more powerful explanation of mathematical learning? For much of this century, absorption theory dominated the field of psychology. Research motivated by absorption theory tends to focus on finding and testing general laws of learning. Such research typically does not study school learning directly. Quite often what are studied are simple and contrived examples of learning, such as memorizing nonsense syllables. Moreover, the research is usually conducted under artificial, well-controlled laboratory conditions—quite often with animals.

Absorption theory and research certainly have value. This approach clearly explains simpler forms of learning, such as the rote memorization of a telephone number, or the formation of a habit, such as how to hold a pencil. However, absorption theory and research have failed to provide a powerful explanation of more complex forms of learning and thinking, such as memorizing meaningful information or problem solving. This approach has, in particular, been unable to provide a sound description of the complexities involved in school learning, like the meaningful learning of the basic combinations or solving word problems.

In recent years, cognitive theory has become the dominant force in psychology, because it seems to provide a more accurate view of learning and thinking in a wide range of circumstances. Cognitive theory and research focus on complex forms of human learning. This approach more adequately explains everyday meaningful learning and problem solving, such as learning how to talk or calculating the best prices while shopping. Indeed, unlike past research efforts, cognitive researchers often focus directly on school learning. As a result, cognitive theory and research can provide a powerful explanation of mathematical learning.

To better appreciate this cognitive perspective, consider the domain of mathematics. What is the nature of mathematical knowledge? What is mathematics? What is mathematical expertise? Commonly, mathematical knowledge is equated with the collection of facts and procedures regarding arithmetic, measurement, and geometry taught in school. Mathematics is often thought of as the science of quantity (arithmetic) and space (geometry). Mathematical expertise is typically associated with a broad and ready knowledge of these domains. Indeed, a common caricature of a mathematician is that of someone who knows many facts and can compute very quickly.

MATHEMATICAL KNOWLEDGE. Order exists in the real world. Heavenly bodies and objects within our grasp do not move capriciously but in regular fashions that can be prescribed. The light that falls upon our eyes is not without regularities but has structure that can be discerned. Mathematical knowledge is an idealized order that can be used to describe, or *model*, the regularities, patterns, and structure of the real world. Mathematical knowledge is a human or mental construction that in part attempts to define or characterize the order we perceive in the world. Consider an elementary example of mathematical knowledge. It is tempting to think of number as a characteristic or property of things in the real world that we uncover. That is, number appears to be an inherent aspect of the physical world that we detect directly. We encounter single objects, pairs of objects, triple instances, and so forth. It could be argued that our senses simply abstract this property from the environment. Indeed, in the sixth century B.C., Pythagoras claimed that the whole numbers beginning with one (1,2,3,4, . . .) were god-given or natural. To this day, this sequence is referred to as the "natural numbers." Many centuries later, the mathematician Leopold Kronecker voiced a similar sentiment: "God created the integer; the rest is the work of man." In other words, number is a natural order that is imposed directly upon our minds. This naturally imposed order serves as the basis for inventing the artificial order that is the rest of mathematics.

Cognitive theory suggests that all mathematical knowledge is a socially agreed upon interpretation or mental invention. Even the so-called natural numbers seem to be a mental construct—an order that we collectively impose upon the environment. Consider "how many" you can readily see in the following picture: **OOO. We can readily see a collection of five. To perform this "feat," we had to artificially treat each object in the picture as part of the same collection or set. That is, we overlooked the very real physical differences between an asterisk and a circle and arbitrarily labeled all the objects as "things." In effect, we artificially defined the collection we saw. We could also agree to interpret the picture as two separate collections or sets: two asterisks and three circles. It is our choice, because number is a subjective, not an objective, reality. Number is an idealized, or *abstract*, model of regularities that we perceive.

Indeed, even 8- or 9-year-olds will point out that numbers do not end, but extend infinitely (Gelman, 1982). However, the inexhaustibility of the number sequence is not a verifiable fact. The infinite cannot be demonstrated through experience; it is an assumption (Dantzig, 1930/1954). Even number, then, is not an order in the real world that we passively note. As with all mathematical knowledge, number is a mental construct—an order actively imposed upon the world. For both mathematician and child, the essence of mathematical knowledge is insight.

MATHEMATICS AND MATHEMATICAL EXPERTISE. Mathematics is more than the finished products of arithmetic and geometry featured in school mathematics. Though mathematics is in part a collection of facts and procedures, it is, at heart, an effort to search for, specify, and use relationships. Indeed, mathematics might better be described as the science of discovering patterns and defining order (Jacobs, 1970). Like meaningful learning by children, then, mathematics is very much an ongoing problem-solving process. It is both accumulated information and a continuing effort to create new knowledge (Davis & Hersh, 1981). Mathematical expertise, therefore, requires insight and problem-solving ability as well as factual knowledge.

EDUCATIONAL IMPLICATIONS: PLANNING FOR MEANINGFUL LEARNING

To make effective decisions concerning mathematics curriculum, instruction, testing, and remediation, educators must accurately take into account the psychology of the child. Educational practice that overlooks how children actually learn mathematics may deter meaningful learning, provoke learning problems, and foment debilitating affect (feelings and beliefs). Explicitly or implicitly, educational decisions are based on one of two theories of learning. For decades, absorption theory has been a principal guiding force for mathematics education. Absorption theory and research have important implications for educators. For example, absorption theory emphasizes the importance of analyzing complex learning tasks in terms of their component parts (task analysis) and then systematically building from basic to complex (i.e., organizing instruction hierarchically).

In recent years, however, cognitive theory has emerged with a more powerful explanation of meaningful learning, including that of direct concern to educators. For example, cognitive theory can help explain the intricacies of meaningfully memorizing the number combinations, learning arithmetic concepts, or acquiring facility with word problems. Cognitive theory proposes that mathematical knowledge is not simply a storehouse of facts and skills that can be readily imposed upon a passive learner. According to this perspective, mathematical knowledge is actively constructed by the child in a manner similar to the problem-solving process used by mathematicians to create new knowledge. In brief, cognitive theory provides a powerful framework for making the practical day-to-day decisions required by mathematics educators. Some general implications for encouraging the active construction of knowledge are described below.

1. *Focus on encouraging the learning of relationships.* Training based on rote memorization has severe limits and drawbacks. Even if possible, it would take enormous time and energy in the form of practice to master even primary-level mathematics by rote. Concentrating on relationships, which summarize whole bodies of information, makes learning primary mathematics a more manageable task. Like adults, children are reluctant to learn meaningless information. Centering on relationship can make learning more meaningful and enjoyable. Children often forget rotely learned information—typically, soon after being tested. Learning relationships is more likely than memorization to promote "retention." Children typically do not see how rotely learned information applies to new but related school tasks or everyday tasks. Learning relationships is more likely to produce "transfer" than memorization is.

2. *Focus on helping children to see connections and change perspectives.* Children's minds are not simply empty vessels that must be filled up with information. Genuine learning involves more than accumulating information. Thus instruction should be more than presenting information and requiring memorization. The most important kinds of learning involve meaningful learning or insight—changes in the way a child thinks about or tries to solve a problem. Meaningful learning involves assimilation and integration of information. To promote meaningful learning, it is important to help children see the connection between instruction and their existing knowledge. Instruction should also attempt to show how bits of knowledge are interrelated.

3. *Plan for meaningful learning to take a long time.* Children often can memorize facts and procedures in short order and on a preset schedule. However, like mastering the basic number combinations, meaningful learning of number, arithmetic, and place value is accomplished gradually insight by insight. Moreover, there is usually a long period of preparation before such reorganizations in thinking occur. Such meaningful learning of mathematics does not adhere to a rigid time schedule. Thus both students and teachers will experience far less frustration if adequate time is allowed for assimilation and integration of knowledge.

4. *Encourage and build upon children's self-invented mathematics.* Children do not passively mimic adults but invent their own means for coping with mathematical tasks. Informal mathematics is a sign of intelligence. To encourage meaningful learning and self-confidence, informal mathematics should be pointed out and praised. When possible, show how a child's invented mathematics and school instruction are connected.

5. *Consider individual readiness.* The state of a child's existing knowledge has relatively little bearing on rote memorization but, in contrast, is critical to meaningful learning. Though insights may occur suddenly, they

do not occur haphazardly. Meaningful learning is unlikely to occur if a child does not have the knowledge to assimilate new instruction. A child must be ready to see connections. Furthermore, as reading instruction is commonly practiced, children should be grouped and taught mathematics on the basis of readiness and need, not age.

6. *Exploit children's natural interest in play*. Many adults disapprove of children's games during school hours. They consider play a frivolous diversion from the real and joyless task of learning: memorizing facts and procedures. In fact, play is children's natural vehicle for exploring and mastering their environment. Games can provide an interesting and meaningful way for learning much of primary-level mathematics. All sorts of games supply opportunities to apply and practice basic arithmetic skills (e.g., Noddings, 1985). Playing games gives children a natural and enjoyable opportunity to make connections and master basic skills and can be invaluable in encouraging either meaningful learning or memorization. Among the books that describe mathematical games for young children are: *A Guide to Teaching Mathematics in the Primary Grades* (Baroody, 1989), *Teaching Mathematics to the Learning Disabled* (Bley & Thornton, 1981), *Learning and Mathematics Games* (Bright, Harvey, & Wheeler, 1985), *Math Activities for Child Involvement* (Dumas & Schminke, 1977), and *Active Learning Experiences for Teaching Elementary School Mathematics* (Lerch, 1981).

CHAPTER 2

Informal Mathematics: The Key Middle Step

Do children come to school with significant mathematical knowledge? What role has concrete experience, especially counting, played in the historical development of mathematical knowledge? What are the nature and extent of children's natural and unschooled mathematics? Why is it important for children to master formal mathematics, and how is initial instruction best accomplished? What are the consequences of overlooking children's mathematics?

THE PRESCHOOLER'S KNOWLEDGE OF MATHEMATICS

Any theoretical understanding of a subject must be grounded in reality and tested by application. In order to keep theory and application solidly connected, case studies will be employed throughout this book. Thus the examination of preschoolers' knowledge begins with a look at a real preschooler.

The Case of Alison

Alison, herself 3½ years old, was celebrating her 2-year-old sister's birthday.

FATHER: Alison, how old is Arianne today?
ALISON: [Holds up two fingers.]
FATHER: How old is Alison?
ALISON: [Holds up three fingers.]
FATHER: How old is Daddy?
ALISON: [Pauses, then holds up *four* fingers.]

The following exchange occurred several weeks later.

FATHER: [Holds up three fingers.] How many fingers are there?
ALISON: [Points with her finger as she counts.] One, two, three.
FATHER: [Holds up two fingers.] How many fingers are there?
ALISON: That's like Eanne [Arianne's age].
FATHER: How many fingers?
ALISON: Two.
FATHER: [Puts out three pennies.] Can you show me with fingers the number of pennies there?
ALISON: [Holds up three fingers, then counts.] One, two, three, four.

Though they are not perfected, this preschooler already has some important mathematical competencies. Alison is quite adept at counting collections of one, two, and often three objects. Indeed, she can even automatically recognize collections of one and two objects as "one" and "two," respectively. Given a small set of objects such as three pennies, she can create a matching model with her fingers. Indeed, for Alison, fingers are a natural medium for expressing mathematical ideas. (She used finger patterns, for example, to represent ages.) Moreover, she appeared to deliberately choose four fingers to represent her dad's age—a representation different from those used to indicate her sister's or her own age. Though inexact, she may have chosen a larger number to indicate an age comparison: Dad is bigger. Is Alison typical of preschoolers? Do most children come to school with basic mathematical skills, such as counting, recognizing, matching, and comparing sets?

Alison's mathematics is based on concrete experiences, such as counting and using fingers. How important are such concrete experiences to children's mathematical development? Alison's mathematics has clear limitations. For example, she accurately counted and recognized very small sets but not larger sets. What are the limitations of children's concrete mathematics? Alison's mathematics is highly practical. For example, she connects finger representations to meaningful events in her life (e.g., two fingers is how old her sister is) and uses them to communicate her ideas and needs. How important is practical need to mathematical development?

Two Views of the Preschooler

Absorption theory assumes that children come to school as blank slates upon which school mathematics can be directly inscribed. Other than perhaps some rotely learned counting skills, preschoolers are seen as devoid of mathematical competence. In fact, the famous associationist theorist E. L. Thorndike (1922) considered young children so mathematically inept that he concluded: "It seems probable that little is gained by using any of the child's time for arithmetic before grade 2, though there are many arithmetic facts that can be learned in grade 1" (p. 198). Moreover, absorption theory suggests that the counting skills children do bring to school are essentially irrelevant or a hindrance to mastering formal mathematics. The acquisition of real mathematical knowledge basically begins afresh with formal training.

Cognitive theory contends that children do not come to school as blank slates. Recent cognitive research shows that before formal schooling begins, most children acquire considerable knowledge about counting, number, and arithmetic. Moreover, this informally acquired knowledge serves as the foundation for understanding and mastering school-taught mathematics. In brief, the roots of mathematical competence extend back to the preschool years, and successful school instruction builds on this informally learned knowledge. To better appreciate the importance of this key building block, let us examine how mathematical knowledge has evolved over the course of human history.

A BRIEF HISTORY OF MATHEMATICS

Concrete Beginnings

BASIC NUMBER SENSE. Humans, like some other species, appear to be endowed with a primitive number sense. We can readily perceive the difference between a set of one and a collection of many or even between a small and a large collection. We can see that something is being added to a collection or taken away. This direct perception can be quite useful in some circumstances but not in others, such as distinguishing a flock of eight birds from one of nine.

CONCRETE TALLYING. To keep track of time and possessions, our prehistoric ancestors devised methods based on matching and one-to-one correspondence. Matching could provide a record of the days, say, from the last full moon: Add one pebble to a pile each night until the full moon appeared again. Likewise, to keep track of a collection of animal skins, a hunter could cut out a notch in a stick or bone for each skin put in a pile. This matching process creates a one-to-one correspondence: Exactly one element in the set of notches for each element in the set of skins. Later, to check if all the skins were still present (if there was still a one-to-one correspondence), the skins could be matched one for one against the notches of the tally stick.

TRACES OF THE PAST. Our languages still bear the traces of prenumber times. For example, English has many ways of expressing "two": *pair*, *couple*, *twin*, *brace*, and so forth. In more primitive times, these terms may have been used to designate the plurality of specific things or categories of things: a pair of eyes, a couple of people, twin trees, a brace of pheasants. Likewise, the numerous terms for "many" (e.g., *multitude*, *lot*, *heap*, *crowd*) once described specific collections larger than two or three (e.g., a *school* of fish; a *flock* of birds). Initially number was simply a quality or feature of a particular object (Churchill, 1961).

Beyond the Purely Concrete

As nomadic hunting and gathering societies gave way to settled communities based on farming and commerce, keeping track of time (e.g., the seasons) and possessions became increasingly important. Thus the need grew for more precise numbering and measuring methods based on counting. *Counting* is the foundation upon which we have built number and arithmetic systems that are pivotal to our advanced civilization. The development of counting, in turn, is intimately tied to our ten fingers. Dantzig (1954, p. 7) concludes:

> It is to his *articulate ten fingers* that man owes his success in calculation. It is these fingers which have taught him to count and thus extend the scope of number indefinitely. Without this device the number technique of man could not have advanced far beyond the rudimentary number sense. And it is reasonable to conjecture that without our fingers the development of number, and consequently that of the exact sciences, to which we owe our material and intellectual progress, would have been hopelessly dwarfed.

Finger counting is the steppingstone for overcoming the limitations of our natural number sense. Where anthropologists have not found evidence of finger counting, perception of number is severely limited (Dantzig, 1954). For example, studies of aborigines in Australia who had not reached the stage of finger counting found but few who could identify four, and none who could distinguish seven. In this natural state, aborigines do not develop basic concepts of quantity and measurement (Dasen, 1972; de Lemos, 1969).

ABSTRACT NUMBER. Counting is probably the means by which our civilization developed an abstract number concept—a concept that makes mathematics possible (Dantzig, 1954). The mathematician Bertrand Russell noted that it may have taken ages before it was recognized that different dualities (e.g., a pair of eyes, a married couple, a brace of pheasants) were all instances of two. Our fingers provide the common ground for designating different concrete dualities as instances of "twoness." The fingers provide ready models for collections of one to ten objects. Two fingers can be held up to indicate, for example, a brace of pheasants or a team of horses. In time, the name for this model collection ("two") could be applied to *any* concrete collection that matched two fingers.

For a long period of history, the terms for "two," "three," and "many" served adequately (Smith, 1923). As the need for greater precision grew, counting became a key tool. Counting places the names of the model collections in an order and provides a convenient alternative to matching for assigning number names. A request could be made directly with a term like *seven* and subsequently fulfilled by counting out seven objects.

CONNECTING THE TWO ASPECTS OF NUMBER. Number has both naming and ordering meanings. The naming, or cardinal, aspect is concerned with how many elements a set contains. Naming a set does not necessarily require counting. As we just saw, a set can be classified as "five," for example, if its elements match up exactly (can be put in one-to-one correspondence) with the elements of a model collection (e.g., the fingers of a hand) called "five." Thus naming sets requires only model collections such as eyes to represent two, a clover leaf for three, legs of a horse for four, and so forth.

The ordering, or relational, aspect of number is tied to counting and concerned with sequencing collections by growing magnitude. Counting provides an ordered succession of words (the number sequence) that can be assigned to increasingly larger collections. To count a collection, a person assigns successive number-sequence terms to each item of the collection until all the items have been assigned a term. The number assigned to the collection specifies the set's relative magnitude. For example, if a collection has been counted and labeled "five," then it is larger than those designated by one, two, three, or four and smaller than those designated by six or its successors.

Finger counting can link the cardinal and relational aspects of number. To represent a collection, say the cardinal number four, a person simply has to raise four fingers simultaneously. To count the same collection, a person raises four fingers in succession. The results of counting are identical to simultaneously raising four fingers (the cardinal representation). Thus, our fingers provide the device for effortlessly passing from one aspect of number to the other (Dantzig, 1930/1954).

The Development of a Base-Ten, Place-Value Numeration System

As societies and economies became more complex, pressure increased to devise recording and calculating systems that could efficiently handle large quantities. To represent a flock of 124 sheep, a one-to-one tally system is very cumbersome. Large-number tasks prompted the idea of grouping, and our 10 fingers provided a natural basis for grouping (Churchill, 1961). For example, as the sheep passed the shepherd, he could keep track on his fingers. When he reached 10, he could represent this count with a pebble. With the hands free again, he could resume the tally. As the number of pebbles accumulated, he may have simplified the tally process further by replacing 10 pebbles with a rock. A rock then represents 10 tens or 100. Because regrouping is based on 10 and factors of 10, the shepherd's tally system is called a base-ten system. If we had 12 fingers, we probably would have regrouped on the basis of 12 and had a base-twelve system today. Our base-ten system is quite simply a "physiological accident" (Dantzig, 1930/1954).

The first known numeral systems evolved about 3500 B.C. and incorporated a base-ten concept (Bunt, Jones, & Bedient, 1976). The cuneiform system of the Sumerians and the hieroglyphic system of the Egyptians both used a collection of strokes to represent the numbers one to nine (see Frame A, Figure 2.1). A group of tens was represented by a special symbol. Later, the Greeks and Romans developed different systems. However, none of these early numeration systems easily lent themselves to arithmetic calculations, as an attempt to add the terms in Frame B of Figure 2.1 will quickly demonstrate.

Though written symbols have been used to represent numbers since prehistoric time, the development of efficient calculation procedures had to await the invention of a *positional* numeration system. In a positional numeration or place-value system, the position of a digit defines its value. For example, in the numeral 37, the 3 is the tens place and hence represents three tens, not three ones. This eliminates the need for special symbols to represent 10 and factors of 10, as with Egyptian hieroglyphics. With a place-value system, 10 digits (0 to 9) can be used to represent any number, even very large numbers, in a compact manner. For example, consider what it would require to represent 9,999,999 in hieroglyphics!

However, positional numeration is a relatively abstract idea, and it was not quickly improvised. The impetus for a positional system was probably the need to make written records of counting-board operations. The counting board depicted in Figure 2.2 utilizes a base-ten model: The right-hand column represents ones, the next column to the left represents groups of

FIGURE 2.1: A Comparison of Different Numeration Systems

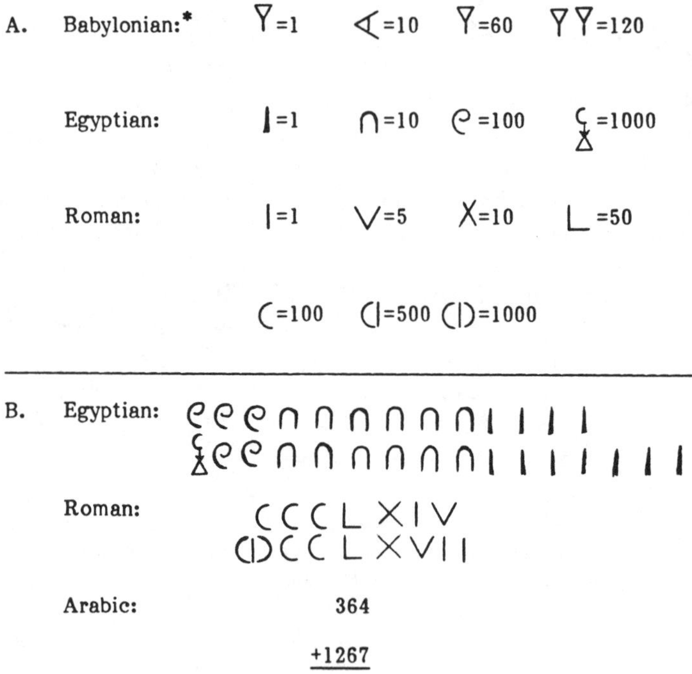

A. Babylonian:* 𒁹=1 𒌋=10 𒁹=60 𒁹𒁹=120

Egyptian: 𓏺=1 𓎆=10 𓍢=100 𓆼=1000

Roman: I=1 V=5 X=10 L=50

C=100 CI=500 CIↃ=1000

B. Egyptian: 𓍢𓍢𓍢𓎆𓎆𓎆𓎆𓎆𓎆𓏺𓏺𓏺𓏺
𓆼𓍢𓍢𓎆𓎆𓎆𓎆𓎆𓎆𓏺𓏺𓏺𓏺𓏺𓏺𓏺

Roman: CCCLXIV
CIↃCCLXVII

Arabic: 364
+1267

*The Babylonian system was adopted from the earlier Sumerian system. Note that the Babylonian numeration began as a base-ten system but then switched to grouping based on 60 and factors of 60. Note that the symbols for 1 and 60 are identical (Bunt et al., 1976). Position was used to indicate value (e.g., 63 was noted as 𒁹 𒁹𒁹𒁹). Like the Babylonian system, Roman numeration was not a purely base-ten system.

tens, and the next column stands for groups of hundreds. Hence, Frame A of Figure 2.2 represents the number four hundred thirty-two (432); Frame B, four hundred two (402).

Counting-board users would have no difficulty with empty columns until they had to make a permanent record of their count. For example, does the record in Frame C of Figure 2.2 represent 42, 402, or 4,002? It seems that zero was invented as the symbol for an empty column so as to avoid this confusion (e.g., Englehardt, Ashlock, & Wiebe, 1984). It appears that zero originally meant empty or blank, not nothing (as in no objects). With the invention of zero, a positional (place-value) numeration system was possible. This made possible arithmetic algorithms that could be learned by nearly everyone. The invention of zero is one of the greatest

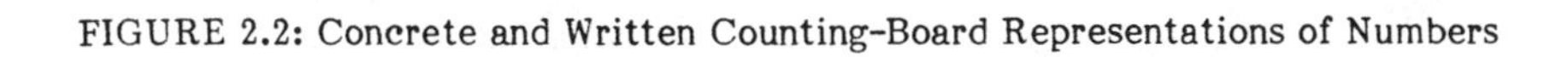

FIGURE 2.2: Concrete and Written Counting-Board Representations of Numbers

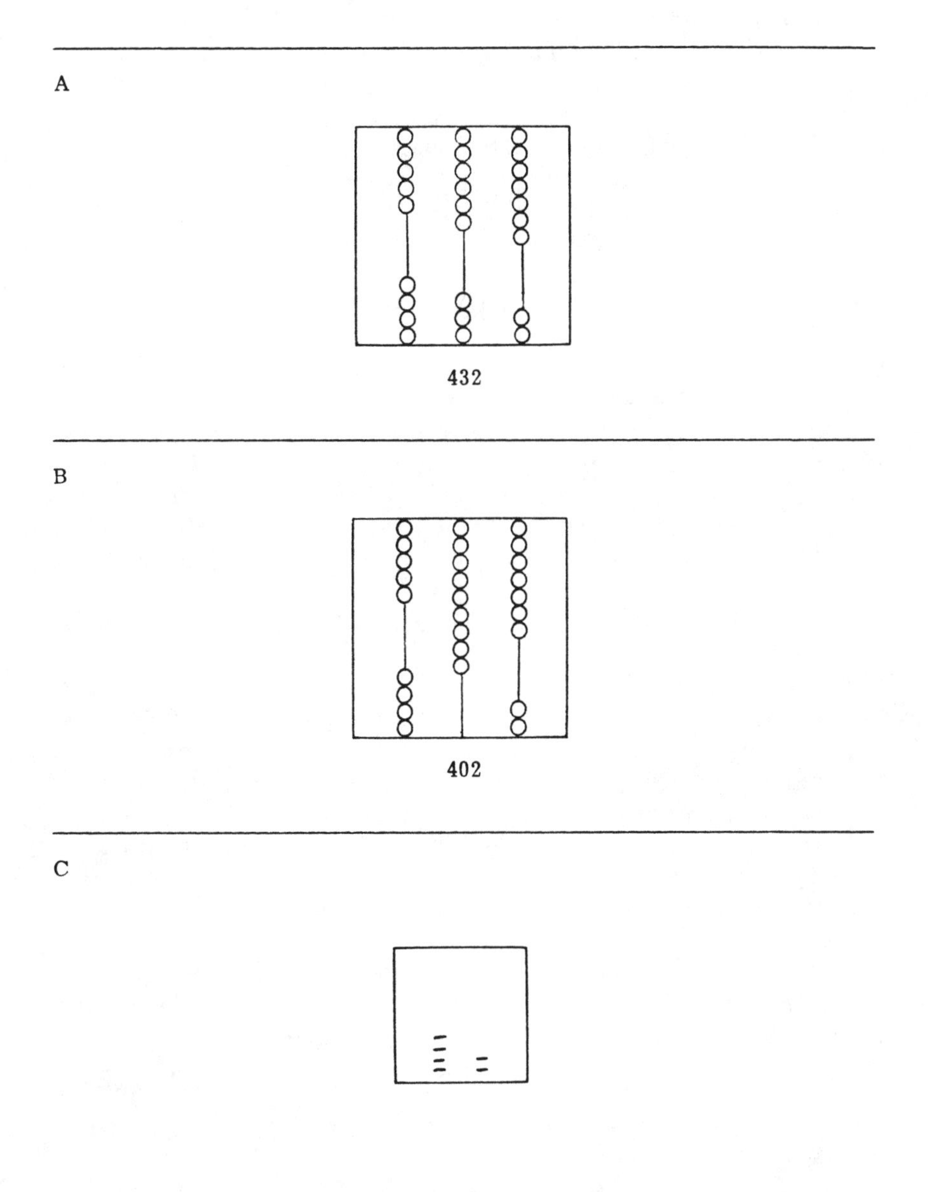

single achievements in human history, for it was a turning point that made possible modern science and commerce (Dantzig, 1954).

In fact, written calculation procedures have been used only for the past 300 years of human history. Only a few hundred years ago, finger counting was the custom in Western Europe. Manuals and universities taught how to do arithmetic computations on the fingers. "The art of using his fingers in counting and in performing the simple operations of arithmetic was then one of the accomplishments of an educated man" (Dantzig, 1954, p. 11).

The Development of Formalized Mathematics

Like the history of number, the history of mathematics in general (see Table 2.1) suggests that intuitive or informal methods and formulations precede and serve as the basis for an exact and formalized mathematics (Kline, 1974). Precise and rigorous deductive proofs (the use of general principles to logically prove points) typically follow inductive insights (the discovery of relationships by examining cases). Basically mathematicians use proofs to check their intuitive or informal insight. Proofs can determine whether or not an insight is logically consistent. Proofs can also demonstrate whether or not an insight applies to an isolated case or a wide range of cases.

A historical perspective indicates that mathematics is actively evolving. Our number and arithmetic systems are the culmination of literally thousands of years of invention and refinement. Mathematical knowledge has been constructed slowly, insight by insight. The knowledge that the average adult in our culture takes for granted was not available several thousand or even several hundred years ago. New approaches were often invented out of practical need and adopted because of practical utility. For example, the Egyptians were forced to invent simple arithmetic and geometry so they could reset the boundary markers of their fields that the Nile flooded each spring (Bunt et al., 1976). Indeed, as Table 2.1 shows, new approaches are often not adopted immediately because they do not "feel right"—because they do not fit into existing patterns of thought.

CHILDREN'S MATHEMATICAL DEVELOPMENT

In many ways, children's mathematical development parallels the historical development of mathematics: Children's imprecise and concrete mathematical knowledge gradually becomes more precise and abstract. It appears that, like primitive humans, infants have some number sense. In

Table 2.1: A Brief History of Mathematical Development

Period	Development
3000–300 B.C.	The Egyptians and Babylonians devise the essential beginnings of mathematics: rudiments of ARITHMETIC (POSITIVE WHOLE NUMBERS & FRACTIONS), ALGEBRA, and GEOMETRY. Results are accepted on a purely EMPIRICAL BASIS. Negative numbers and 0 are not known.
600–300 B.C.	Classical Greece is the first civilization in which mathematics flourishes. Classical Greeks are the first to conceive of DEDUCTIVE MATHEMATICS. Euclid's *Elements* (geometry proofs) are the product of 300 years of intuitive exploration and error.
	Alexandrian Greeks, Hindus, and Arabs devise and use IRRATIONAL NUMBERS (e.g., $\sqrt{2}$), which are gradually accepted because of their utility (e.g., $\sqrt{2}$ = the diagonal of a square with sides of 1).
A.D. 600	Hindus introduce NEGATIVE NUMBERS, which are not accepted for a thousand years because of the lack of intuitive support. For example, the great mathematicians Descartes and Fermat refused to work with negative numbers.
Circa A.D. 700	Hindus either invent or adopt a symbol for "zero" to indicate an empty or blank column. Arabs adopt Indian numeration and, after hundreds of years, Arabic numerals come into common use.
Circa A.D. 1540	COMPLEX NUMBERS appear (e.g., $\sqrt{-1}$) but are not accepted for about 200 years.
A.D. 1650–1725	Newton and Leibniz create the CALCULUS. Three editions of Newton's *The Mathematical Principles of Natural Philosophy* each presented a different explanation of the basic concept (the derivative). Leibniz's first paper was called an "enigma" rather than an explanation. Despite its vague and even incorrect foundation, calculus found wide application via intuitive approaches.
Late 1900s	The logical foundations of the number system, algebra, and analysis (the calculus and its extensions) are erected.

See Kline (1974, pp. 41-47) and Bunt et al. (1976, pp. 226-230) for a more detailed description.

time, preschoolers build on their intuitive mathematics and develop a wide range of competencies. Recapitulating history, children's unschooled or *informal mathematics* grows out of practical concerns and concrete experience. As it did historically, counting plays a central role in the development of this informal knowledge. In turn, children's informal knowledge paves the way for the *formal mathematics* taught in school. And, paralleling cultural history, mastering positional numeration and the calculational algorithms based on this concept is a giant step for children. Indeed, because it often does not fit into their existing pattern of thought, children do not immediately accept and learn the formal mathematics taught in school.

Intuitive Knowledge

NATURAL NUMBER SENSE. It has long been thought that infants are essentially devoid of mathematical thought. William James once characterized the infant's world as a blooming, buzzing confusion. Yet recent research (e.g., Starkey & Cooper, 1980; Starkey, Spelke, & Gelman, in press) suggests that even 6-month-old infants can distinguish between sets of one and two, two and three, and three and four.

How can psychologists tell that infants have this basic number sense? To see if an infant can discriminate between sets of different numbers, a psychologist presents an infant with, say, a picture of three objects (e.g., Starkey & Cooper, 1980). Interested by the novel stimulus, the baby stares at the picture. After repeated displays of three, however, the novelty wears off and attentiveness drops off. At this point, the psychologist introduces a set of four (or two) objects. If the child does not notice the difference, the inattentiveness should continue. However, infants tend to become attentive again, suggesting that they do notice the difference.

Is the infant really attending to changes in number? In the example above, infants gradually become bored with "threeness" even though different objects are introduced or the position of the three objects is changed. The type or arrangement of the objects does not affect attentiveness. Indeed, after viewing examples of three objects, infants are less interested in hearing a sequence of three sounds than in hearing a sequence of two or four sounds. It appears that it is *threeness* that they no longer find interesting. Apparently, infants have a primitive enumerating or matching process that permits them to distinguish among small sets of objects.

An infant's number sense is limited in scope and precision. Infants cannot distinguish between larger sets, such as four and five. Moreover, though infants seem to treat, say, threeness and fourness as different, this does not necessarily mean they know that four is *more than* three. That is, even though infants distinguish between small classes of numbers, they may not order them successively in terms of magnitude.

INTUITIVE NOTIONS OF MAGNITUDE AND EQUIVALENCE. Nevertheless, children's basic number sense provides the foundation for mathematical development. Preschoolers build on this number sense and develop more sophisticated intuitive knowledge. It is from the concrete experience of direct perception that children begin to understand such notions as relative magnitude. Concretely there is an obvious difference between one and larger collections (von Glasersfeld, 1982). A child can pick up, say, one block with one hand. Two blocks require both hands or two successive efforts with one hand. Three blocks cannot be picked up simultaneously with two hands. Though these differences may seem trivial to an adult, they are of paramount concern to a young child at play and provide another concrete basis for distinguishing among and ordering one, two, and many.

By the time they are toddlers, children not only distinguish among sets of different size but can make rough magnitude comparisons. At about the age of 2 years, children learn words to express mathematical relationships (Wagner & Walters, 1982) that can be attached to their concrete experiences. They can note "same," "different," and "more." For example, Alfred Binet (1969), the father of modern intelligence tests, asked his 4-year-old daughter, Madeleine, to compare the size of sets, like the two shown in Figure 2.3. Though Madeleine could count only to three, she was fairly accurate in pointing out sets with more.

Repeated testing demonstrated that Madeleine's intuitive judgments of more were based on perceptual cues, such as how much area the sets covered (Ginsburg, 1982). Madeleine intuitively picked as more the set that was more spread out. This was often successful because the larger set usually took up more space. Other perceptual cues, such as length, also can provide a basis for intuitive evaluations. Under many circumstances, the longer of two rows does have more objects.

Recent research confirms Binet's finding. When asked to choose which of a pair of sets has more, 3-year-olds, disadvantaged preschoolers, and young children from nonliterate cultures can do so quickly without counting (Baroody & Ginsburg, 1982b). Nearly all children entering school should be able to distinguish between and label as "more" the larger of two *obviously* different sets. (Correctly using "less" is much more difficult and may not be learned before school.) A child unable to use "more" in this intuitive manner is at considerable educational risk.

However, because children base their judgments on appearance, their magnitude comparisons are sometimes incorrect. Though appearance often accurately reflects amount, perceptual cues, such as area and length,

FIGURE 2.3: Example Items of a Test of More

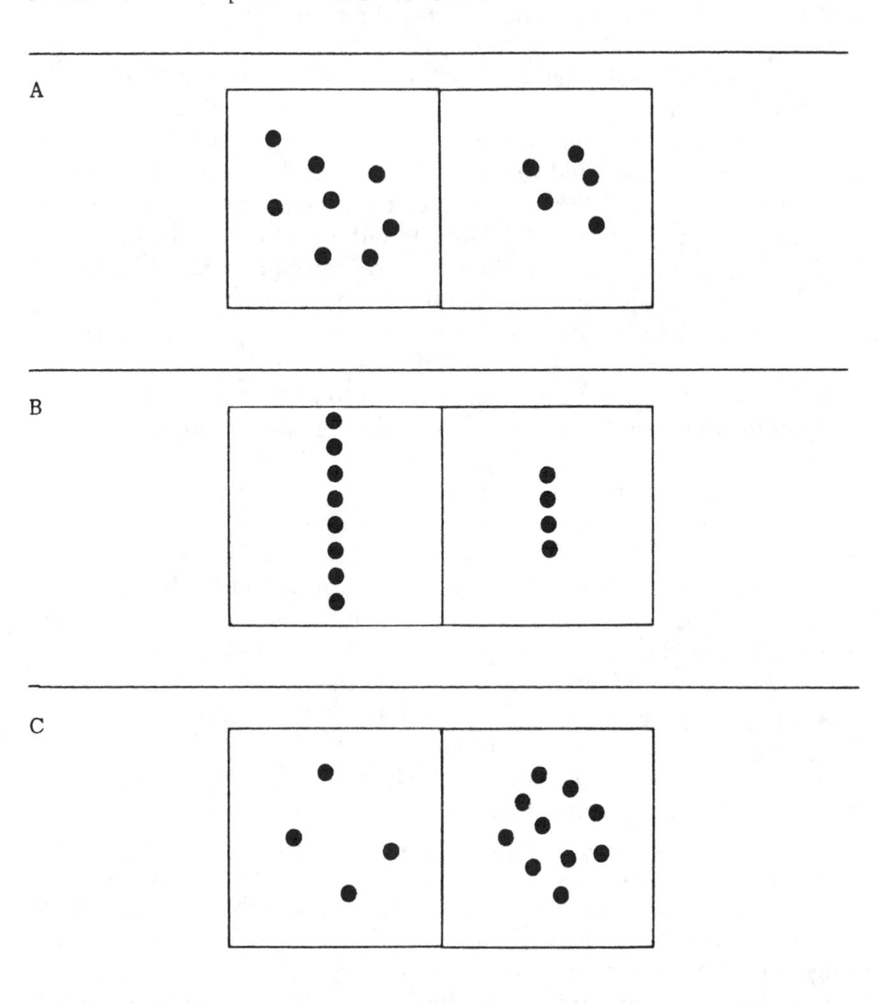

are not always accurate indicators of number. For example, two plates of candies can cover the same area but be unequal in number because the candies on one plate are more tightly packed together. On the other hand, two plates of candies may have the same number but take up different amounts of space because the candies on one plate are more densely packed!

The number-conservation task (e.g., Piaget, 1965) dramatically demonstrates the limitations of children's intuitive knowledge. First, the equivalence of two sets is established by matching. The tester sets out a row of, say, seven white blocks, and the child is asked to put out the same number of, say, blue blocks. The child is encouraged to match one blue block to each white. Once a one-to-one correspondence has been established (see Frame A of Figure 2.4), the child is asked to confirm that the two rows have the same number of objects. Because the length clue provides an accurate basis for judging relative amount, even 3-year-olds will agree that the two rows have the same number.

Then the appearance of one of the sets is changed to evaluate whether or not the child continues to believe that the two sets are equivalent in number. While the child watches, one row is lengthened or shortened. For example, Frame B of Figure 2.4 shows that the blue row has been lengthened. After the length transformation, the child is asked again if the two rows have the same number. Because length cues no longer accurately reflect number, children who rely on appearance to judge amount are now misled. Indeed, young children will insist that the longer row now has more! They really seem to believe that the two sets of different lengths are nonequivalent. Piaget (1965) called this phenomenon nonconservation because the child does not maintain (conserve) the initial equivalence relationship over a (number-irrelevant) transformation in appearance. Clearly, children's intuitive understanding of magnitude and equivalence is imprecise.

INTUITIVE NOTIONS OF ADDITION AND SUBTRACTION. A number sense also permits children to recognize that a collection has been altered. Children recognize very early that adding an object to a collection makes "more"; removing an object diminishes a collection. In one study (Brush, 1978), preschoolers were shown two containers. Screens were placed before the containers to hide them from the view of the child. Using a matching process, an equivalent number of objects were placed in each container: As an object was placed in one container, an object was also placed in the second container. After the child indicated that the two hidden containers contained the same number of objects, the child observed an object either being added to or taken out of one of the containers. The children had no difficulty recognizing that the addition or subtraction of objects changed the amount in one container and, as a result, changed the equivalence relationship. For example, children readily identified as "more" the container to which an object was added. It seems that preschoolers already have an intuitive basis for understanding addition and subtraction.

However, intuitive arithmetic is limited to obvious changes. If the containers initially contained unequal amounts, intuitive arithmetic fails. For example, if five objects are initially placed in one container and nine in

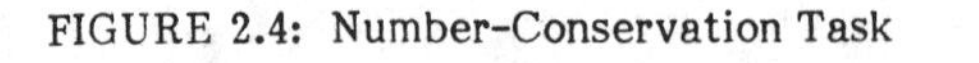
FIGURE 2.4: Number-Conservation Task

A. Time 1: Child creates equivalent sets by matching.

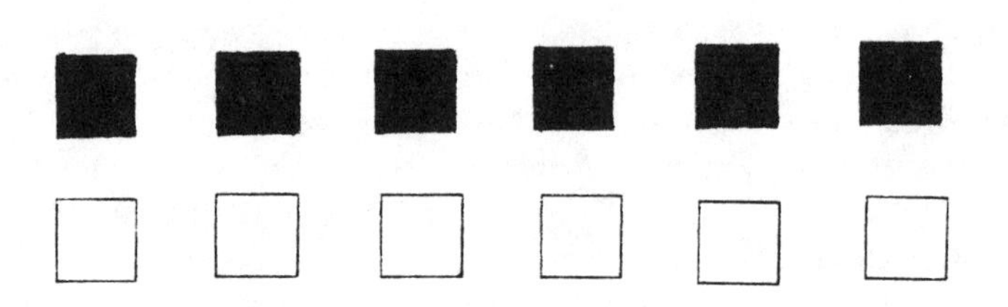

B. Time 2: The appearance of one row is changed: The row of blue blocks is spread out.

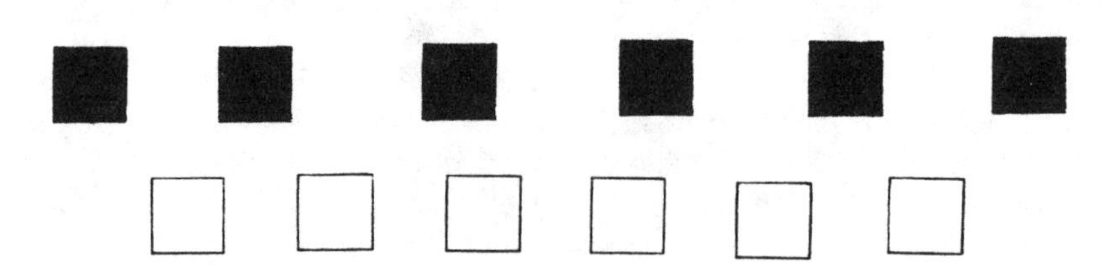

another, children will correctly identify as "more" the container with nine. But then if two objects are added to the container with nine and four are added to the container with five, children identify the latter as having more. To young children, 5 + 4 is "more than" 9 + 2 because they saw more added to that container. Clearly, intuitive arithmetic is imprecise.

Informal Knowledge

A PRACTICAL EXTENSION. Children find that intuitive knowledge is simply not sufficient to cope with quantitative tasks. Thus they increasingly rely on more precise and reliable tools: number and counting. Indeed, soon after they start speaking, children begin to learn number words. At about 2 years, they use "two" to designate all pluralities: two or more objects (Wagner & Walters, 1982). By about $2^1/_2$ years, children begin to use "three" to designate "many" (more than two objects). Like Alison, many 3-year-old children use "one," "two," and "three" reliably and choose a term larger than three (e.g., "four") to indicate "many." By labeling collections with numbers, children have a precise device for determining "same," "different," or "more." Preschoolers even discover that counting can be used to determine exactly the effects of adding or subtracting at least small numbers from a collection.

Counting thus builds on and greatly supplements intuitive knowledge. For example, counting provides a common label ("three") for triples of different arrangement and different objects that children, as early as 6 months, see as equivalent. By using direct perception in conjunction with their counting, children discover that number labels like "three" are not tied to the appearances of sets or objects and are useful in specifying equivalent sets. In time, this insight provides the basis for using number labels like "seven" or "nineteen" for identifying as equivalent, sets that cannot be seen as equivalent. Counting provides children the bridge between concrete but limited direct perception and abstract but general mathematical ideas. It is counting that puts abstract number and simple arithmetic within the reach of the young child.

LIMITATIONS. Though informal mathematics represents a fundamentally important elaboration over intuitive mathematics, it also has practical limitations. Counting and informal arithmetic become increasingly less useful as numbers become larger. The time and mental effort required to count or informally calculate become enormous and eventually prohibitive. As numbers become larger, informal methods become increasingly prone to error. Indeed, children may not be able to use informal procedures with larger numbers at all. Moreover, informal methods provide an immediate solution but no long-term records.

Formal Knowledge

The written, symbolic mathematics taught in school overcomes the limitations of informal mathematics. Formal mathematics can free children from the confines of their relatively concrete mathematics. Written symbols supply a means of noting and dealing with larger numbers. Written procedures furnish efficient devices for arithmetic calculations with large numbers. Moreover, written symbols and expressions can provide clear and permanent records that can greatly extend memory capacity.

It is essential that children learn base-ten/place-value concepts. To cope with larger quantities, it is important to think in terms of ones, tens, hundreds, and so forth (Payne & Rathmell, 1975). Thinking in groups of ten and factors of ten gives children flexibility and facility in dealing with a wide range of mathematical tasks, including ordering (comparing) larger numbers and mental arithmetic with multidigit terms. Place value provides the underlying rationale for many basic skills, such as writing multidigit numerals and renaming (carrying or borrowing) (Resnick, 1982, 1983). In brief, formal mathematics enables children to think in more abstract and powerful ways and to deal efficiently with large-number problems.

Though formal mathematics can greatly extend a child's capabilities, it involves learning new skills and concepts that initially may seem foreign and difficult to many children. Children become accustomed to thinking of numbers and arithmetic in terms of counting. A numeral such as 14 is viewed as 14 units or as 13 units and 1 more. Positional notation is not readily apparent to young children. As it did historically, a grasp of positional notation evolves slowly for children. Thus it may take children a while to see, say, 14 as one ten and four ones. The idea of zero as a placeholder (as representing an empty column) may take a very long time to develop. Indeed, many children may cling to informal or concrete methods well after positional notation and renaming algorithms are introduced.

EDUCATIONAL IMPLICATIONS: BUILDING ON INFORMAL KNOWLEDGE

Cognitive theory points out that children just entering school are not simply empty vessels that must be filled up with knowledge. Most children—including those from low-income families—come to school with a good deal of informal mathematical knowledge (Russell & Ginsburg, 1984). Indeed, many special education children have at least some informal knowledge (Baroody, 1983a; Baroody & Ginsburg, 1984; Baroody & Snyder, 1983). Preschoolers learn a good deal of informal mathematics from family, peers, TV, and games before entering school.

Children's informal mathematics is the crucial middle step between children's limited and imprecise intuitive knowledge based on direct perception and the powerful and precise mathematics based on abstract symbols taught in school. As it did historically, the practical and relatively concrete experience of counting provides children a basis for acquiring number and arithmetic competencies. Because learning involves building upon previous knowledge, informal knowledge plays a key role in the meaningful learning of formal mathematics. Because learning is an active process of assimilating new information to what is known, informal knowledge is a crucial basis for understanding and learning school-taught mathematics. Cognitive research indicates that, regardless of how mathematical skills, symbols, and concepts are introduced in school, children tend to interpret and deal with formal mathematics in terms of their informal mathematics (Hiebert, 1984). Informal mathematics, then, is the foundation for "mastering the basics" and successfully tackling more advanced mathematics. Two key educational implications of this view are described below.

1. *Formal instruction should build on children's informal mathematical knowledge.* It is essential that educational planning take into consideration children's informal mathematical knowledge. Teachers need to exploit informal strengths to make formal instruction meaningful and interesting. In addition to increasing the likelihood of successful school learning, exploiting informal strengths can have important affective consequences. The principle of relating formal instruction to informal knowledge is applicable to the whole range of primary-level topics, from mastering the basic number combinations to learning place-value concepts and procedures such as computing with renaming. We will also see that this principle applies to children with a wide range of abilities, including the learning-disabled and mentally handicapped.

2. *Gaps between informal knowledge and formal instruction frequently account for learning difficulties.* When such formal instruction is introduced too quickly and does not build upon children's existing informal knowledge, the result is rote learning, learning problems, and/or destructive beliefs. Unable to connect formal mathematics to anything meaningful, many children just mechanically memorize and use school-taught mathematics. Some children fail at even memorizing facts and skills. Many

children remain uninterested in the subject, acquire a distaste for the topic, or even come to dread mathematics.

A gap between formal instruction and children's informal knowledge is especially likely to cause learning difficulties with relatively abstract base-ten/place-value skills and concepts. As a result, many children have difficulty grasping positional notation and experience problems with related skills, such as borrowing. Many have trouble with base-ten representation and fail to develop efficient skills for handling larger numbers. Special education children in particular may have great difficulty in negotiating the transition from informal arithmetic based on counting to formal arithmetic based on positional notation.

CHAPTER 3

Two Views of Curriculum and Instruction

How do absorption and cognitive theories differ in terms of the aims and methods of mathematics education? How does each define the role of the teacher? What view is traditional instructional practice based upon? Does such an approach promote intelligent learning and use of mathematics? What was the aim of the New Math? Was this reform effort successful or not and why? Should primary instruction focus on teaching arithmetic or teaching mathematics? How important is active involvement, like discovery learning, in the elementary years?

EDUCATIONAL EFFECTIVENESS AND THEORETICAL PERSPECTIVE

Genuine learning results from an *interaction* of *external* factors, such as the nature of the subject matter and teaching practices, and *internal* factors, such as a child's learning ability and interests (Dewey, 1963). When school mathematics and the psychology of the child mesh, children learn and use mathematics in a meaningful and even enthusiastic manner. When subject matter and teaching practices are not suitable, schooling may fail to encourage real learning and may even create learning problems. Consider the following case, in which a father set out to teach his 22-month-old twins to count objects.

> FATHER: Dad will show you what to do. Here 1, 2 [pointing in turn to each large dot on a card]. Can you do that, Alexi?
> ALEXI: 3, 4 [pointing correctly to each dot in turn].
> FATHER: Alison, can you go 1, 2 [pointing in turn to each dot] for Dad?
> ALISON: 3 [pointing to the last dot counted by her father].
> FATHER: Can you go 1, 2 [again demonstrating the counting process]?
> ALISON: 3, 3 [slapping her hand over the card].

FATHER: Alexi, watch carefully. 1, 2, 3 [counting each dot of a three-dot set]. Can you do that?
ALEXI: 1, 2, 3, 4, 5 [accompanied by four points toward the dots].
FATHER: Alison, can you go 1, 2, 3 [modeling the object-counting procedure again]?
ALISON: 3 [pointing twice to each of the last two dots]!
FATHER: Ready, 1, 2, 3 [again demonstrating the object-counting process]. Can Alexi do that?
ALEXI: 2, 3, 4, 5 [pointing on or near each dot in turn].
FATHER: Alison, can you go 1, 2, 3 [accompanied by pointing]?
ALISON: T [waving her fingers over the dots]!

Because Alexi and Alison were not ready for or saw little purpose in the instruction, the lesson had little or no impact on them. Their father's experience parallels that of many teachers, who regularly find that their students just do not listen. This raises an important question: Is direct instruction the best vehicle, or even a viable one, for educating young children?

Whether or not an educator properly takes into account internal factors depends upon his or her theory of learning. How or even whether direct instruction should be used is shaped by one's beliefs about learning. According to absorption theory, education involves inculcating knowledge from without. According to cognitive theory, education involves encouraging understanding and critical thinking from within. It follows that these views define the nature and content of the curriculum, the goals and methods of instruction, and the teacher's role quite differently.

Absorption Theory

THE PRIMARY CURRICULUM. Absorption theorists regard primary mathematics as a body of basic and socially necessary facts and skills. More specifically, the curriculum is seen as a collection of facts and procedures regarding arithmetic, geometry, and everyday applications: basic arithmetic facts, computational procedures, and definitions (e.g., + means "plus or add"; addition is the union of two sets); facts and skills pertaining to two- and three-dimensional objects (e.g., the names of shapes and procedures for determining perimeter and area); and practical facts and skills (e.g., the names of coins and how to exchange coins, how to use a ruler or read a clock, using shapes to recognize common road signs). In effect, absorption theory treats mathematics as a ready-made *product* that schooling must help children absorb (Romberg, 1984).

INSTRUCTION. According to absorption theory, the aim of instruction is to help children acquire the facts and procedures contained in the curriculum. Indeed, mastery of computational facts and skills is often seen as the main instructional goal of primary mathematics education (Fey, 1979). The recent "back-to-basics" movement reinforces this view (Lindquist, 1984).

According to absorption theory, the methods of instruction are clear. Because children are simply uninformed, direct instruction is the most efficient way to introduce facts and skills to children. Thus instruction based on absorption theory typically relies heavily on verbal and textbook explanations and relatively abstract written symbolism. For instance, instruction on the basic addition combinations might begin with the teacher pointing out that 7 plus 3 equals 10. Verbal instruction is reinforced with a written demonstration at the chalkboard and by printed examples in the pupils' mathematics textbooks. The children may begin to cement the bond between $7 + 3$ and 10 by imitation: chanting along with the teacher that 7 plus 3 is 10 or copying the teacher's example on the chalkboard. Practice then serves to implant the fact firmly in memory. For example, the association is then firmly imprinted by verbal drills and numerous seatwork, worksheet, and homework assignments.

Because it is assumed that children merely have to be receptive, group instruction is deemed appropriate. Because the capacity to accumulate information is considered basically uniform (at least among typical children), instruction can proceed at one pace. Except for unusual cases, then, individualized instruction is not essential.

TEACHER'S ROLE. Absorption theory suggests a well-defined and straightforward role for the teacher: to transmit information. The teacher must orchestrate all that goes on in the classroom — presentations, demonstrations, assignments, and rewards and punishments — with that goal in mind. As long as a topic is presented clearly and children are attentive, acquisition of basic facts and skills should proceed smoothly toward mastery; a good teacher's curriculum guide should provide an effective basis for planning daily instruction in all but exceptional cases.

Cognitive Theory

PRIMARY CURRICULUM. Cognitive theorists regard primary mathematics as a system of fundamental ideas and ways of approaching mathematical problems. In this view, "the test of learning is not mere mechanical facility in 'figuring' " (Brownell, 1935, p. 19). The primary goal of school mathematics should be to cultivate comprehension and intelligent use of mathematical relations and principles. In effect, cognitive theory views school mathematics as a *process* of promoting more sophisticated mathematical understanding, reasoning, and problem solving.

INSTRUCTION. In a cognitive approach, the aim of instruction is to help children construct a more accurate representation of mathematics and develop more mature thinking patterns. Teaching mathematics is essentially a process of translating mathematics into a form children can comprehend, providing experiences that enable children to discover relationships and construct meanings, and creating opportunities to develop and exercise mathematical reasoning and problem-solving abilities.

Cognitive theory suggests that instruction that actively involves children is the best means for helping them construct mathematical understanding and develop more mature thinking patterns. There are numerous ways to actively involve pupils in learning. Using games and concrete manipulatives is often suggested. Another frequently mentioned technique is structured or guided discovery learning, a prescribed activity that helps children to discover a relationship. One vehicle, which is often overlooked, is small group discussions (Cobb, 1985a), where children might share questions, insights, and strategies. Another underused vehicle is tutoring: Teaching another (perhaps a younger) child can be a powerful incentive for learning. The give-and-take of questioning children about their work can thoughtfully engage pupils. Indeed, even a direct explanation can sometimes be thought-provoking *if* it comes at a time when the child is ready. In brief, actively involving children is essential to meaningful learning and can take many forms.

Actively or meaningfully involving pupils in instruction is difficult to accomplish with mass education. Because of individual differences in readiness, meaningful learning can be quite uneven. Instruction then should be tailored to individual needs.

TEACHER'S ROLE. Cognitive theory suggests that teaching is essentially a problem-solving process that requires great flexibility and knowledge. In this view, the teacher acts as an intermediary—someone who helps external factors and internal factors to mesh. Therefore, to be an effective teacher requires knowing the subject matter, teaching techniques, *and* the child.

Because every situation and child are different, teachers must continuously make "educated guesses" about how to proceed. Furthermore, they need to check or evaluate how effective their decisions have been. In effect, teachers must be hypothesis makers and hypothesis testers. Example 3.1

EXAMPLE 3.1

A Case of Educational Problem Solving

Worksheet 4 DAVID

✓	1 + 1 = 0	✓	3 + 2 = 1
✓	2 - 1 = 3	✓	4 - 1 = 5
✓	2 + 2 = 0	✓	1 + 3 = 2
C	4 - 2 = 2	C	4 - 3 = 1
✓	1 + 4 = 3	✓	2 + 1 = 1
C	5 - 4 = 1	✓	3 - 1 = 4

Scientific Method	Teacher's Problem-Solving Process
1. RECOGNIZE THE PROBLEM:	"David could not do his math seatwork."
2. FORMULATE A HYPOTHESIS:	"Maybe he can't read numerals and he's just guessing."
3. TEST THE HYPOTHESIS:	The teacher asks David to read the numerals from 1 to 9 on the number list above the chalkboard. David does so perfectly.
4. DRAW A CONCLUSION:	"Reading the numerals is not the problem."
2A. REVISE THE HYPOTHESIS:	"Perhaps he can't do simple addition or subtraction. Maybe he doesn't understand the concepts of addition and subtraction."
3A. TEST THE HYPOTHESIS:	"Though usually incorrect, his answers aren't random. For 1 + 1, it looks as though he subtracted correctly to get his answer. For 2 - 1, it appears that he added correctly to get his answer." The teacher checks further by asking David some word problems (e.g., "If you had one penny and I gave you one more, how many would you have all together?") David figures out the word problems correctly.
4A. DRAW A CONCLUSION:	"Given the consistency of his error pattern, it seems that David does know how to add and subtract. Moreover, given his success on simple word problems, it appears that he understands the basic concepts of addition and subtraction."

(*continued*)

EXAMPLE 3.1 (continued)

Scientific Method	Teacher's Problem-Solving Process
2B. REVISE THE HYPOTHESIS:	"Perhaps his seatwork difficulty was due to the fact that he does not know the plus and minus symbols."
3B. TEST THE HYPOTHESIS:	The teacher asks David what the plus symbol on his worksheet means. David explains that it means either "all together" or "take away" but that he is not sure.
4B. CONCLUSION:	"I need to help David distinguish between the plus and minus symbols and what operation each represents."

This example was contributed by Sharon C. Baroody.

illustrates this practical problem-solving process and compares it with an idealized description of the scientific method (mathematical or scientific problem solving).

The first-grade teacher described in Example 3.1 noticed that David was having a problem with his basic arithmetic assignment and formulated a hypothesis about the difficulty. She tested this educated guess, but the results were not consistent with her hypothesis. She concluded that her hunch was incorrect and that she needed a new explanation. After formulating and evaluating two additional hypotheses, the teacher identified the specific deficiency that was causing the child's problem. Once she had figured out the basis—the underlying psychological cause—of the problem, David's teacher was in a position to effectively remedy the boy's problem. Like mathematicians or scientists, educators must be active problem solvers, continually making and testing hypotheses.

ARITHMETIC OR MATHEMATICS?

A Focus on Arithmetic

Since early in this century, absorption theory has been the dominant model for mathematics education in the United States (Romberg, 1982). Instruction based on absorption theory focuses on arithmetic: encouraging memorization of arithmetic facts and skills. Children are required to memorize number facts, definitions, computational procedures, measurement skills, and so forth. Given limited resources and large classes, teaching and drilling facts and skills are more manageable than encouraging conceptual knowledge and reasoning ability. For example, teaching the step-by-step borrowing algorithm for multidigit subtraction is easier than building up the network of relationships that constitutes knowledge of place value. Moreover, knowledge of facts and skills is easier to observe and test than conceptual knowledge or reasoning ability. Because cultivating and assessing mathematical understanding, reasoning, and problem solving are difficult, mass education focuses on teaching and testing arithmetic facts and skills.

However, mathematics taught in a highly didactic and lockstep manner overlooks the psychology of the child. Instruction based on absorption theory typically overlooks the importance of children's informal knowledge, readiness to learn, and natural interests. By only briefly using physical models to introduce arithmetic and then quickly jumping to drilling number facts and arithmetic algorithms, most instruction disregards children's need for a prolonged period of informal figuring (Carpenter & Moser, 1984). Mass education is often blind to individual differences in personal interests and informal readiness to learn school mathematics.

Even efforts to make instruction concrete often fall short. For example, textbooks commonly illustrate base-ten grouping with bundles of 10 or 100 sticks (e.g., see Figure 3.1). Many children are just baffled by and ignore such diagrams. Even if they think to count the sticks in the bundle, the task

FIGURE 3.1: Typical Textbook Illustrations for Base-Ten Notions

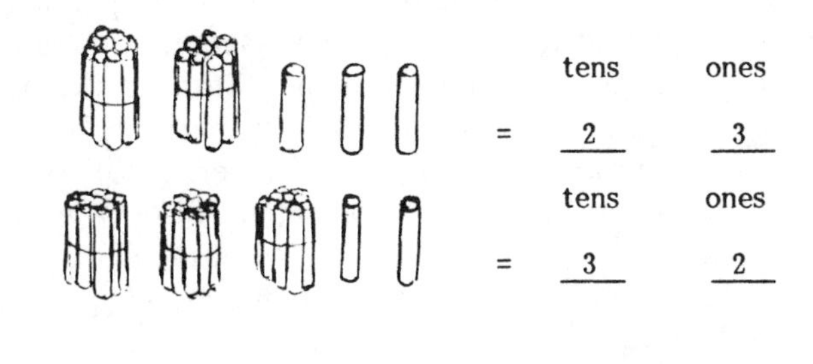

is of very little interest to a child. Moreover, pressure to move on to other assignments and topics does not give the child an opportunity to reflect on the exercise.

When presented with abstract, complicated, and uninteresting instruction children ignore, misconstrue, forget, or—at best—rotely memorize the new information. Like Alison and Alexi, schoolchildren may simply not listen to instruction for which they are not ready. They may apply their own incorrect interpretation to unfamiliar instruction (e.g., a child might interpret the newly introduced symbol for multiplication [×] as referring to the familiar operation of addition). Children—like anyone—are prone to forget information that is not personally meaningful. Even if children do rotely remember basics, it does not guarantee intelligent use of the knowledge. In effect, many children master arithmetic at the expense of learning mathematics. These problems are made worse when drill and practice are uninteresting and meaningless. Mass instruction too often actually discourages meaningful learning, thinking, and problem solving.

Mechanical Behavior

When purposefully and meaningfully engaged in a task, children spontaneously search out and use relationships and monitor and adjust their actions. For example, when given a sequence of commuted problems, such as $9 + 7 = \_$ and $7 + 9 = \_$ or $12 + 5 = \_$ and $5 + 12 = \_$, many first graders quickly recognize that it is unnecessary to compute the sum of the second problem of a commuted pair (Baroody, Ginsburg, & Waxman, 1983). Exploiting relationships, like commutativity, to shortcut effort is an example of intelligent problem solving (Wertheimer, 1945). When actively engaged in a mathematical task, children tend to behave intelligently (see Example 3.2).

However, many children do school mathematics perfunctorily. Their mechanical behavior is evidenced in inefficient problem solving, inconsistencies, and nonsensical answers. For example, John Holt (1964) observed fifth graders who laboriously used the standard algorithm to calculate the sum of $256 + 328$ even after just calculating and recording the sum of $256 + 327$. Holt also observed fifth graders who were sure *both* $245 + 179 = 424$ and $245 + 179 = 524$ were correct, despite the fact that the addends were identical and the equations were written side by side on the board. As the work below shows, some elementary schoolchildren subtract only to get differences that are *larger* than the minuend or initial amount (Hiebert, 1984).

EXAMPLE 3.2

The Case of Tracey

When children are actively engaged in tasks, they check their work and correct errors without prompting. Consider the case of Tracey, an 11-year-old Down's Syndrome child with an IQ of 53. Asked to count out eight pegs from a pile of pegs, Tracey correctly counted out five items but then incorrectly labeled the sixth item "6, 7." Apparently sensing that she had made a mistake, Tracey spontaneously checked the number of pegs taken: "1, 2, 3, 4, 5 [pause], 6." Apparently realizing that she had forgotten how many she was supposed to take, Tracey did not proceed with the task but indicated that she needed a reminder. She recounted the six pegs and then added two more as she counted, "7, 8."

The videotaped interviews for this case study were conducted by Cathy A. Mason.

$$\begin{array}{r}107\\-\ 34\\\hline 133\end{array}\qquad\begin{array}{r}125\\-\ 83\\\hline 162\end{array}\qquad\begin{array}{r}117\\-\ 48\\\hline 131\end{array}$$

Children frequently are undisturbed by inefficiencies, inconsistencies, or errors that—to an adult—seem needless. Why? Sometimes children simply do not have the knowledge necessary to solve a problem more effectively or to monitor their work accurately. Fast-paced, verbally based instruction is often not assimilated by children. Therefore, they often do not have a basis for tackling problems or checking their work. However, sometimes children simply fail to connect what they do know to what they are doing. That is, sometimes children simply fail to use their resources to guide their problem-solving efforts or monitor their work.

Instruction based on absorption theory often leads to blind procedure following, which bypasses existing knowledge and interferes with intelligent problem solving. For example, first-, second-, and third-grade children were given a sequence of problems $6+7$, $6+8$, $6+9$, . . . , $6+15$ to see if they would shortcut their computational efforts by simply adding one to the previously figured sum (Baroody et al., 1983). Not surprisingly, the second graders used the $N+1$ shortcut far more frequently than the first graders. Curiously, though, the second graders also far outperformed the third graders, who generally persisted in *computing* each successive problem. Likewise, another study (Bisanz, LeFevre, Scott, & Champion, 1984) found that 6-year-olds tended to use an inverse principle to solve problems such as $5+3-3=\_$ (i.e., recognized that, in effect, the $+3$ and -3

canceled each other), while 9-year-olds tended to laboriously compute the answers (e.g., computed 5 + 3 and then 8 – 3). In both studies, elementary schoolchildren did not use principles to reason out answers quickly, as younger subjects had, but instead laboriously computed each answer.

Because they have learned to follow procedures blindly, children often fail to use the knowledge at their disposal to monitor their work. Hiebert and Wearne (1984) found that many sixth graders incorrectly added decimals such as 3.71 + 1.2 (answered 3.83 rather than 4.91), because they lined the numerals up on the right (as in the addition of whole numbers) rather than lining up the decimal points. These same students were asked to predict the most likely error for the problem 3.71 + 1.2 by a student who was not very good at decimals. Some predicted that the poor student would make the very error they had made a few minutes earlier! According to one child, the reason for this mistake is " 'cause they're used to regular pluses and they're not really used to doing decimals." Thus, despite their knowledge, these children failed to check and correct their decimal computations.

Reform Efforts: The New Math

The New Math was an attempt to reform traditional mathematics instruction in the United States. After the Soviets launched *Sputnik* in 1957, the U.S. educational system was heavily criticized for relying too heavily on rote memorization. Traditional mathematics was seen as incomprehensible, dull, irrelevant, and even frightening for all too many children (Kline, 1974). The reformers felt that "teaching the structure of the subject" would foster meaningful learning (e.g., Bruner, 1963). More specifically, the New Math was born of the idea that, if the precise and logical nature of mathematics could be revealed to children, they would understand and even enjoy mathematics. If it was taught precisely and logically, children would not have to rely on rote memorization to learn mathematics.

The New Math adopted as its basis the set theory and deductive-reasoning approach of theoretical mathematics. Children are first expected to learn definitions, such as addition or joining sets: "One set joined to another is the set consisting of all objects which belong to either set" (School Mathematics Study Group, 1965, p. 109). Children are also required to learn axioms (basic relationships), such as the associative axiom: $(a + b) + c = a + (b + c)$. Then, using definitions and axioms, children can prove conclusions (engage in deductive reasoning). Take, for instance, the problem 4 + 12 = __. (a) By definition, 12 equals 10 + 2; therefore, 4 + 12 = 4 + (10 + 2). (b) The associative law stipulates that 4 + (10 + 2) = (4 + 2) + 10.

(c) By definition 4 + 2 equals 6 and 10 + 6 equals 16. Note that such an approach has the advantage of making explicit relationships (e.g., the associative law) and concepts (e.g., the proof makes clear that the 1 in the positional notation 12 actually stands for 10).

However, does the New Math really help children to understand and appreciate mathematics better than the traditional approach it was supposed to replace? Like the traditional curriculum, the New Math did not adequately take into account the psychology of the child. Though it was an effort to highlight the structure of the subject matter (a key external factor), the New Math overlooked how children think and learn (internal factors). Because the instruction does not involve their original thinking, children often derive little understanding or insight from the definitions, axioms, and proofs (Kline, 1974). Precise formulations thrust upon the unready are just a meaningless jumble of words. As a result, many children simply resort to rote memorization—the very kind of learning the New Math was designed to avoid. Moreover, this highly formal approach hides the practical and informal origins of mathematics. "[The child] is led to believe that mathematics is created by geniuses who start with axioms and reason directly from the axioms to the [conclusions]. The student, unable to function in this manner, feels humbled and baffled" (Kline, 1974, p.54). Thus, to children, the New Math is no less confusing, boring, irrelevant, and threatening than traditional mathematics.

EDUCATIONAL IMPLICATIONS: PLANNING BALANCED AND APPROPRIATE INSTRUCTION

Effective educational planning considers how external factors, such as the organization of curriculum content and teaching practices, interact with internal factors, such as the child's psychological readiness for learning a particular topic. The aim of instruction based on absorption theory is memorization of arithmetic facts and procedures; the primary methods are direct instruction and practice. Because their role is transmitting information, teachers concentrate on how to complete the prescribed sequence of instruction, how to present lessons, and what work should be assigned. That is, planning focuses on the important external factors of curriculum organization and teaching practices. However, because it overlooks internal factors, instruction based on absorption theory frequently creates learning difficulties or promotes mechanical behavior.

The aim of the New Math was to overcome the limitations of a rote-memorization approach by highlighting the structure of mathematics. In organizing a curriculum and teaching a subject, it is important to underscore relationships. Nevertheless, the New Math did not adequately consider internal factors and, as a result, was no more meaningful to children than more traditional instruction.

According to a cognitive approach, the aim of mathematics education is the cultivation of mathematical understanding and thinking. As they are ready, children should be helped to construct a more elaborate, abstract, and precise idea of mathematical relationships. They should be helped to develop more logical reasoning and problem-solving abilities. This can only be accomplished if a knowledgeable, skilled, and flexible teacher carefully matches instruction to the level, needs, and interests of students. A cognitive approach, then, emphasizes the importance of taking into account internal factors and their interaction with external factors. This approach suggests that teaching, like learning or mathematics itself, is at heart a problem-solving process. Some implications of a cognitive view are delineated below.

1. *Instruction should encourage* both *arithmetic mastery and mathematical thinking*. It is, of course, important for children to master basic arithmetic knowledge. Knowledge of specific facts is needed for higher-order learning, reasoning, and problem solving. Mastery of basic skills and applications is important for further studies and a productive adult life. However, arithmetic should not be taught at the expense of mathematical thinking. Now, more than ever, it is essential that children develop mathematical reasoning and problem-solving competencies. Because work habits that facilitate or hinder reasoning and problem solving develop quite early, it is important that thinking be encouraged from the beginning of school.

2. *Instruction should employ a wide variety of techniques that actively involve the child in learning*. Highly verbal or formal instructional approaches too frequently do not actively engage children's minds and, as a result, fail to inspire understanding and thinking. Moreover, an overemphasis on quickly memorizing facts and procedures to keep in step with the rest of the class frequently induces only mechanical learning and use of mathematics. Instruction needs to actively engage children's minds. This can be accomplished through "child-centered" activities, such as math games. Note, though, that games or other activities that directly involve children do not guarantee that a child's mind is actively engaged. Activities have to be matched to the child's readiness and needs. Games and other activities need to be supervised so that a child receives guidance and correction when needed.

"Passive" techniques, such as teacher explanations, questions, and demonstrations, can play an important role in primary mathematics instruction if used skillfully and carefully. For example, an effective lecture can help children see connections (relationships, implications, and applications) as well as provide basic information. *If* carefully matched to a child's readiness, need, and interest, passive techniques can be instructive and interesting. Though sometimes overused and misused, passive teaching techniques do not necessarily lead to rote learning.

3. *Structured discovery learning should be an important ingredient in primary-level mathematics.* Guided discovery learning is a highly suitable means for bringing together external and internal factors to teach primary-level mathematics. Primary-level mathematics affords numerous opportunities to discover important mathematical relationships. If given the opportunity, children—even special education children—can be guided to important mathematical discoveries. The discovery of relationships underlies meaningful learning and encourages thinking skills. Moreover, because it exploits children's natural curiosity, discovery learning promotes excitement about mathematics and learning in general. Thus children should be given the opportunity to search for and summarize relationships on a regular basis.

4. *Instruction should be moderately novel.* When it does not fit into their existing pattern of thought, children may tune out instruction and become bored and restless. (Children may react likewise if instruction is obvious and unchallenging.) To actively engage a child's attention, intelligence, and curiosity, instruction should be somewhat unfamiliar. This is why teaching primary-level concepts and skills in terms of children's informal mathematics is so important. This teaching principle also underscores the importance of matching instruction (external factors) to individual need (internal factors). An instructional activity that some children might find boring because it is either too unfamiliar or too familiar might be enthralling for those at an appropriate level of readiness. This principle highlights the complex problem-solving task that confronts teachers.

CHAPTER 4

Different Approaches to Evaluation and Remediation

How do evaluation and remedial instruction based on absorption theory and cognitive theory differ? Why is it important to look beyond the correctness of children's answers? Do standard tests provide adequate information for educational diagnosis? What is the nature of effective educational diagnosis, and what kinds of information should it provide for effective remedial planning?

THE NEED FOR SPECIFIC EVALUATION: THE CASE OF ADAM

The school district had amassed considerable information about Adam, an 11-year-old with severe mathematical learning difficulties. Neurological examinations indicated organic brain dysfunction; medication controlled seizures. School records indicated that grade-equivalent scores on the KeyMath test at the end of the third and fourth grades were 2.4 and 2.7, respectively. A recently administered Stanford Achievement Test yielded a total math score in the 2nd percentile (concepts, 2nd percentile; computation, 1st percentile; applications, 18th percentile). A recent WISC-R IQ test revealed a normal verbal IQ (100) and a depressed performance IQ (77)—producing a full-scale score of 87. He appeared to have age-appropriate vocabulary and abstraction ability.

In spite of the large amount of test information, there was no clear direction on how to teach Adam. Years of remedial efforts—including assignment to self-contained learning-disabilities classes—had produced negligible progress. Teachers, parents, and Adam were terribly frustrated. Why had all the testing not provided better direction? Why had his remedial program not been more successful? Was it his learning disability or an ineffective diagnostic and remedial approach that prevented Adam from progressing beyond a second-grade level?

TWO VIEWS OF EVALUATION AND REMEDIATION

The nature of evaluation and remediation is shaped explicitly or implicitly by a theory of learning. Because a learning theory defines the goals of education, it dictates what is important to examine. Thus it determines how children are tested and how learning difficulties are diagnosed. A theory of learning also molds the aims and methods of remedial efforts. Absorption and cognitive theories suggest very different approaches to testing, diagnosis, and remediation.

Absorption Theory

WHAT NEEDS TO BE EXAMINED? According to absorption theory, the focus of school mathematics is the mastery of facts and skills, and so the point of evaluation is to assess whether or not this mastery has been achieved. *Testing* focuses on how much a child has learned. A test score provides a general index of a child's level of achievement. *Diagnosis* aims at obtaining more specific information on mastery: what facts and skills the child has and has not mastered.

In effect, evaluation based on absorption theory focuses on external performance: what the child *produces*. Testing gauges mastery in terms of the number of correct answers (*accuracy*). Sometimes testing also mea-mastery in terms of accuracy and quickness (*efficiency*). Diagnosis gauges mastery in terms of which items are answered accurately or efficiently. Testing and diagnosis based on absorption theory both focus on product: how accurately or efficiently a child responds.

THE MEANING OF ERRORS. According to absorption theory, errors simply indicate a deficiency: a lack of mastery. A child is inaccurate or inefficient because facts and skills have not been adequately infused. For example, a mistaken number fact implies an insufficiently strong bond with the correct answer. Mistakes in arithmetic calculations suggest weaknesses in the series of associations that make up the algorithm. Sometimes these deficiencies are attributed to external factors, such as insufficient practice (Thorndike, 1922). Frequently, these deficiencies are blamed on general internal factors, such as student carelessness (lack of interest or laziness), inattentiveness, inability, or low mathematical aptitude.

REMEDIAL EFFORTS. According to absorption theory, the remedy for inadequate mastery is clear. The materials must be reviewed with the child, and the child must practice the material further. If evaluation shows that a child is too far behind his or her peers, the child must repeat a grade or be assigned to a special education class for intensive remedial instruction. Intensive remedial efforts are often reduced to assigning children two or three times the normal amount of practice (Moyer & Moyer, 1985). In theory, a huge amount of practice is necessary for "overlearning." That is, because children with learning difficulties are slower to absorb knowledge or more readily forget knowledge, these children must overpractice a fact or skill to firmly embed it in memory.

Cognitive Theory

WHAT NEEDS TO BE EXAMINED? In a cognitive view, because school should focus on meaningful learning and thinking ability as well as mastery of basic information, evaluation is necessarily an intricate task. In addition to gauging fact and skill achievement, evaluation needs to address such questions as: What concepts or understanding does a child have? Does the child approach problems in a reasonable manner or not? To address these issues, it is not enough for evaluation to focus on external behavior—to determine how much and what a child has mastered. For purposes of diagnosis especially, it is essential to gauge the child's internal state or underlying knowledge. Thus it is important to determine how and why—the *process* by which—a child arrived at an answer (Bruner, 1966; Glaser, 1981). What conceptions and misconceptions did the child bring to the task? What strategy did the child employ to solve the problem?

Examining process has a number of important advantages over evaluation based exclusively on noting the correctness of answers. To begin with, it provides a richer portrait of a child's true competence. For example, Felicia, a kindergartner, was screened on arithmetic ability. She could correctly compute the sums to problems, such as 3 + 2 = __, 2 + 4 = __, and 3 + 5 = __. Such information is interesting in itself and useful in grouping and instructional planning. However, actually observing Felicia's efforts provided even more interesting and useful data. Typically the girl started with one, counted up to the larger number, and then counted on the smaller term (e.g., 2 + 4 = "1, 2, 3, 4 [pause], 5, 6") (Baroody, 1984a). By examining process, it was clear *how* the child accurately computed sums. Such information can provide clearer direction for instructional planning. Felicia could already choose the larger of two single-digit numbers. Thus time and effort did not have to be wasted teaching her this basic skill. Instead of unnecessary and boring instruction, new challenges could be planned. In particular, Felicia might be encouraged to start with the larger number instead of beginning with one each time (e.g., "2 + 4 = *4*; 5, 6").

Examining process also provides a more accurate picture of a child's competence. Evaluation that examines product exclusively may overestimate a child's competence. An example of "false success" is illustrated in Example 4.1. Second, just tallying the number correct may underestimate a child's competence. Many things (e.g., a momentary distraction, unclear wording, tiredness) may cause children to miss an item even when they have the skill or understanding supposedly assessed by the item. An exam-

EXAMPLE 4.1

A Case of False Success

On the following exercise, George got 7 of 10 problems correct. Because 70% was considered passing, George was given credit for the lesson objective: "The child will correctly identify equivalent and inequivalent expressions."

Exercise 29B
Equivalence and Inequivalence

Name: George

Indicate whether the equals and unequals signs have been used correctly. If the statement is true, write T. If the number sentence is incorrect, write F.

a. T $5 + 3 = 8$ C

b. F $7 + 2 = 10$ C

c. T $1 + 3 \neq 4$ ✓

d. T $8 = 7 + 1$ C

e. T $6 + 2 \neq 2 + 6$ ✓

f. T $5 = 5$ C

g. T $3 + 1 \neq 5 - 1$ ✓

h. T $2 + 2 + 2 = 6$ C

i. T $3 = 4 - 1$ C

j. T $9 \neq 8$ C

Had George really mastered the objective? With a choice of only two answers, George stood a 50-50 chance of getting any particular question right just by guessing. On average, guessing should permit a pupil to get about 5 of the 10 correct. In this case George was somewhat luckier than most, but lucky nevertheless.

ple of a false failure—what psychologists call a "performance failure"—is described in Example 4.2. Finally, examining process is especially important in the case of a "competence failure" (when children miss test items because they truly lack mastery). A focus on performance overlooks invaluable information needed to diagnose incomplete or inaccurate understanding or reasoning and to design an effective remedial plan.

THE MEANING OF ERRORS. According to cognitive theory, errors do not merely indicate a knowledge deficiency. They can reveal what knowledge the student brought to bear on the problem and how the child tried to cope with the problem. Errors provide a window to children's internal thought processes and indicate how well these thought processes match up with a learning task.

Consider Peter's addition test results in Figure 4.1. If a teacher considered only accuracy, the situation looks bleak indeed. After a whole unit on

EXAMPLE 4.2

Examining Process to Uncover Understanding

A diagnostic teacher administered the following achievement-test item to a child:

In the picture below, which duck is FIRST?

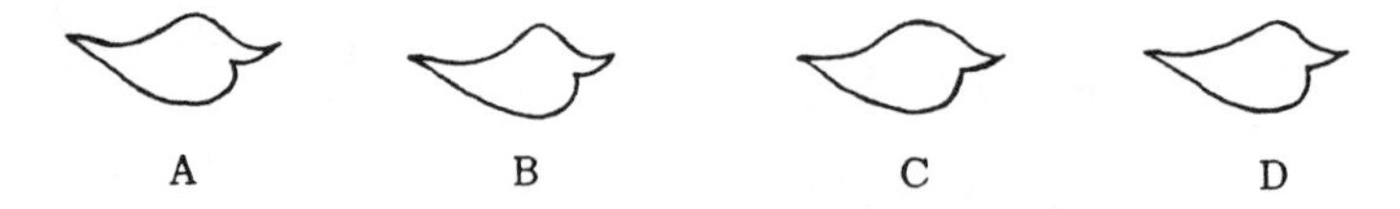

A B C D

The child responded, "A." Was the child right or wrong and why? To answer these questions, the tester had to uncover the child's method of solution. The ordinal numbers (first, second, third, fourth, etc.) define position or order in regard to a reference point. The orientation of the ducks in the picture implied that the reference point was the right-hand side of the page. Thus, technically the correct response is Duck D. The child who responded "A" may simply have been guessing, may not have realized that ordinal numbers imply a particular reference point (a common gap in knowledge), or overlooked the implied reference point and used her own (the left-hand side of the page). By asking a probe question ("How did you figure that out?"), the diagnostic teacher found that the child actually understood the concept despite "missing" the test item. The child explained, "A is the first duck from the left."

This case study was reported to me by Barbara S. Allardice, mathematics coordinator of the Learning Development Center at the Rochester Institute of Technology.

FIGURE 4.1: Examples of Errors

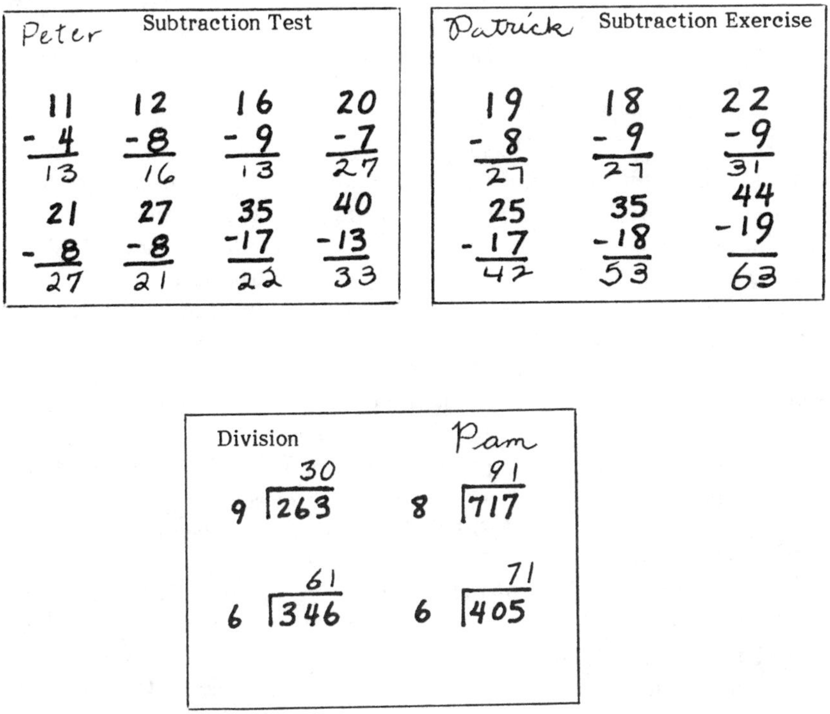

Peter — Subtraction Test

11	12	16	20
- 4	-8	- 9	-7
13	16	13	27
21	27	35	40
- 8	- 8	-17	-13
27	21	22	33

Patrick — Subtraction Exercise

19	18	22
- 8	- 9	-9
27	27	31
25	35	44
- 17	-18	-19
42	53	63

Division — Pam

$9\overline{)263}$ = 30	$8\overline{)717}$ = 91
$6\overline{)346}$ = 61	$6\overline{)405}$ = 71

renaming, Peter did not get a single problem correct! Are Peter's answers haphazard—an indication that he was extremely careless or was simply guessing? In examining his answers, is there anything you notice? It appears that Peter could correctly obtain the difference between single-digit terms but always subtracted the smaller from the larger. Clearly, then, Peter's errors were not random. *Systematic errors*, such as the small-from-large "bug," are the product of a methodical though incorrect strategy (Buswell & Judd, 1925; Brown & Burton 1978). (In computer terminology, a bug is an error in a computer program that produces systematically incorrect answers.)

When children, such as Peter, encounter a task they do not entirely understand, they frequently try to make do with what they do know. Because borrowing did not make sense to Peter, he failed to learn the borrowing algorithm. When he encountered problems requiring borrowing, he fell back on the familiar procedure of subtracting the smaller digit from the larger. In effect, he invented his own incorrect method for coping with an unfamiliar task. The small-from-large bug may be reinforced by the teaching dictum often heard in primary classrooms when subtraction is first introduced: "Always subtract the smaller from the larger." Some children may continue to follow this advice even in situations when it is no longer appropriate. In any case, systematic error strategies are based on the child's previous learning.

Examine the work of Patrick in Figure 4.1. What can you discern about his errors? Patrick appears to simply add rather than subtract. When presented with an unfamiliar or unmanageable task, children often resort to using a previously learned (but inappropriate) procedure. Perhaps Patrick is unfamiliar with the subtraction symbols or does not have a method that can generate subtraction answers quickly enough to finish his assignments in the allotted time. As a result, he handles the task in a way that has worked for him in the past—by adding.

Pam's method (see Figure 4.1) is not as obvious. Pam's problem can be traced to the fact that she has learned only part of the school-taught algorithm.[1] For $263 \div 9$, for instance, she has been taught to look for the product of 9 that is closest to but not more than 26 ($9 \times 2 = 18$). However, Pam did not learn or forgot the qualifying phrase, "but not more than." Her partially correct learning of the algorithm leads her to look for the product closest to 26 ($9 \times 3 = 27$).

In brief, when children encounter tasks they are not ready for or have not been given enough time to master, they often resort to invented, inappropriate, or partially incorrect procedures that produce systematic errors. Systematic errors, then, can provide an invaluable signal that instruction is out of synchrony with the psychology of the child—that external and internal factors are not meshing.

REMEDIAL EFFORTS. Cognitive theory suggests that review and practice may not be effective means for remedying learning difficulties. Indeed, such an approach may actually make problems worse. When a child is having learning difficulties, a teacher needs to consider how instruction can be adjusted so that it harmonizes with the child. The nature of the adjustment depends on considering the underlying thought processes of the child.

Consider the case of Helene. The report on Helene indicated that she was having trouble with fractions. The diagnostic teacher[2] tried several explanations and demonstrations to help Helene understand fractions bet-

1. This case study was reported to me by Barbara S. Allardice, mathematics coordinator of the Learning Development Center at the Rochester Institute of Technology.

2. This case study was reported to me by Karen Combs, diagnostic-teaching instructor at the Learning Development Center, Rochester Institute of Technology.

ter, but no improvement was made. After a frustrating session for everyone, the diagnostic teacher examined a sample of Helene's work for clues concerning the child's thinking (see Figure 4.2). To determine the top number or numerator, the girl simply counted the number of shaded parts. To determine the bottom number or denominator, the girl just counted the total number of parts. So in the case of the first problem, she noted that there was one shaded part and three parts altogether for an answer of $^1/_3$. The child did not realize that a fraction represents a part of a total made up of *equal* parts. Thus, in the first problem, a correct answer is $^1/_2$ (or $^2/_4$) because there are one of two (or two of four) equal-sized pieces of the pie shaded. Recent research (e.g., Behr, Lesh, Post, & Silver, 1983) suggests that Helene's misconception is common among children.

Once she recognized the misconception underlying Helene's difficulty, the diagnostic teacher was in a position to remedy the difficulty. The teacher began by taking out a candy bar and saying, "I'll give you half of this candy bar." The teacher then cut the candy bar into two unequal parts and offered Helene the smaller portion. Helene protested that she had not received her fair share. This demonstration provided the basis for a discus-

FIGURE 4.2: Helene's Fraction Work

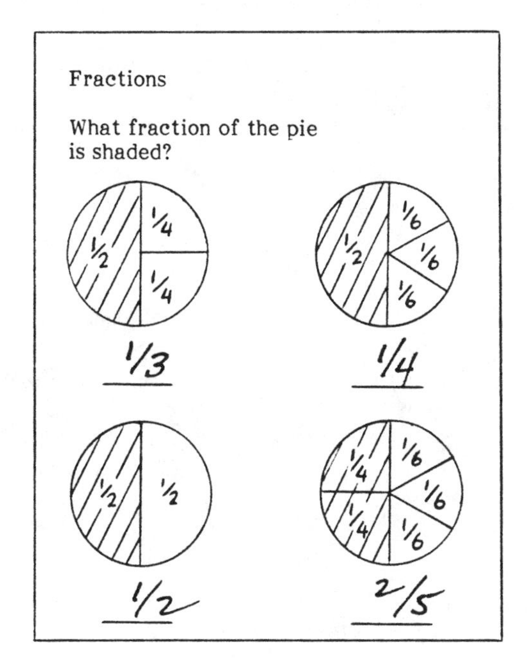

sion on the necessity of taking into account equal-sized parts when writing fractions.

Cognitive theory suggests that giving children more of what did not work in the first place is not likely to help remedy learning difficulties. It is discouraging to be reinstructed in the incomprehensible and to redo assignments that seem pointless or insurmountable. Re-presenting the same lesson to Helene that she failed to understand in the first place did not help. In fact, it only served to make Helene feel worse. Moreover, additional drill would have served little purpose other than to frustrate both pupil and teacher. Indeed, further practice may simply have reinforced Helene's error, making it all the harder for her to unlearn her incorrect procedure and to learn the correct procedure later (Moyer & Moyer, 1985).

The reteaching-and-massive-doses-of-practice routine contributed to Adam's difficulties. Though he was assigned to a special education class, he repeatedly received the same kind of instruction. Abstract and verbal instruction was no more meaningful the second, third, or fourth time he heard it. The volumes of worksheet exercises, textbook assignments, and dittos served only to make Adam more rigid and apathetic. Because of the constant, uniform practice of a single skill, Adam had a tendency to use the same skill throughout an assignment. For example, if addition problems were embedded among subtraction problems, he would subtract regardless of the sign. Moreover, after years of meaningless instruction, uninspiring practice, and humiliating failure, Adam had given up caring about mathematics. By trying to understand *why* children make errors, a teacher is more likely to formulate effective remedial instruction. Clearly, diagnostic and remedial efforts need to focus on the underlying causes of difficulties.

STANDARD TESTS

Standard tests are evaluation instruments that are administered in a highly prescribed (standard) manner. Many but not all standard tests are designed to be administered to groups of children. Standard tests are used for both testing and diagnosis. However, are such tests useful in uncovering underlying processes, particularly the causes of difficulty? Do standard tests provide the diagnostic information needed to plan remedial efforts?

Norm-Referenced Tests

Some commercially available standard tests, such as the Iowa Test of Basic Skills and the California Achievement Test, evaluate *how much* a child has mastered in comparison to children nationally. Such standard tests are called norm-referenced tests (NRTs) because a child's performance is compared to national norms. Such tests are devised primarily to rank children in terms of general achievement. They provide a formal comparison in terms of percentiles, grade-equivalent scores, or other "standardized" scores.

NRTs have a number of uses. Because they are based on national norms, NRTs can serve to compare a child's, a class's or a school system's relative accomplishments. NRTs are also a basic means for classifying and grouping children. The results of NRTs, for example, are the basis for defining a child as learning disabled or mentally handicapped and assigning the child special education services. For instance, NRTs showed that Adam had basically normal intelligence but was more than two grade levels behind in mathematics achievement. This justified his classification as learning disabled and assignment to a special education class in his school. NRTs are widely used because of the ease of administration and the availability of norms for judging relative success.

NRTs are useful for categorization but not diagnosis. Though NRTs have become a common fixture in educational evaluation, they do not provide the diagnostic information needed for effective educational planning. NRTs yield a single score that summarizes a child's achievement. Some NRTs, like the KeyMath (Connolly, Nachtman, & Pritchett, 1971/1976), yield subtest scores, which can give a rough indication of children's ability in various areas of work. For example, the KeyMath indicated that Adam did not have grade-level computational skills and was especially deficient in the area of concepts. However, the general results of NRTs do not provide a detailed view of a child's strengths and weaknesses – a profile of what specific skills or concepts a child is and is not capable of. NRTs are simply not designed to be diagnostic instruments.

Criterion-Referenced Tests

Some standard tests are potentially more useful in diagnosis because they can be designed to gauge specific competencies. Criterion-referenced tests (CRTs) are designed to evaluate a child's achievement in terms of a set of instructional objectives. That is, a child's performance is evaluated in a prescribed manner in terms of a preestablished standard (criterion) for mastery (e.g., given ten two-digit addition problems that do not involve carrying, a child will accurately compute the sum of at least nine). Thus, the objective (two-digit addition without carrying) and the criterion for success (90% accuracy) are clearly spelled out. If children meet the criterion, they are given credit for the specific proficiency.

Because they detail strengths and weaknesses, such tests provide clear direction for instructional planning. CRTs are especially useful in evaluating a subject domain like mathematics where specific skills can be delineated and where complex skills clearly build on more basic skills. Like NRTs though, CRTs are designed to focus on product rather than process.

The KeyMath

The KeyMath Diagnostic Arithmetic test illustrates the diagnostic limitations of standard tests. The KeyMath is widely used, especially with exceptional children, for evaluating elementary mathematics ability and as a basis for developing instructional/remedial plans (Kratochwill & Demuth, 1976). The KeyMath is designed to provide diagnostic information on four levels, such that each succeeding level provides more specific information (Connolly et al., 1971/1976). *Total test performance*, in the form of a grade-equivalent score, provides general placement information. *Area performance* indicates general ability in content (knowledge and concepts), operations (adding, subtracting, multiplying, dividing, mental computing, and numerical reasoning), and applications (e.g., solving word or missing element problems, measuring). *Subtest performance* scores for each of the 14 subtests are also available. Finally, *item performance*, which can be analyzed with the help of behavioral objectives listed in the appendix, indicates mastery of specific skills.

Unfortunately, the design of the KeyMath greatly limits the diagnostic information it can provide. Total test performance (the grade equivalent) of the KeyMath is a normative score and does not provide any diagnostic information (Underhill, Uprichard, & Heddens, 1980). Even the item-performance level, with a criterion-referenced format, fails to provide specific and useful diagnostic information. There is only one test item per objective, which greatly limits interpretation of results (Underhill et al., 1980). For example, if a child is right on Item 4-11 ("This number is 19. What number comes right before this number?"), is it safe to conclude that the child is competent with larger numbers such as 47 or 99, as implied by the objective ("Given a two-digit number, identifies the two-digit number that precedes it")? On the other hand, if a child is wrong, does this imply an inability with all teen numbers or even single-digit numerals? What is the source of the difficulty? Is the deficiency due to unfamiliarity with the number sequence or the term *before*?

Moreover, many objectives are not measured effectively. Some items are prone to overestimate ability. For instance, for Item D-1, the child is shown a picture with one match on the left and two on the right and asked, "One match and two matches are how many matches?" The item is supposed to gauge whether or not the child can do simple addition with concrete objects. Yet a child does not need any understanding of addition to respond successfully. All the child needs to do is interpret the task as a request to count (as in Items such as A-1, A-4, and A-8). Success on Item J-5 (picture of three books and five books and the question, "Henry has 3 books and Jim has 5 books. If they divide the books equally, how many books will each boy have?") is intended to measure an ability to compute an average ("Given a problem requiring addition and division, solves for the average"). However, a child—without knowledge of arithmetic or the procedure for computing an average—can simply look at the picture and see that if one book is taken from the set of five and placed with the set of three, the sets will be equal and each will contain four.

Some items may underestimate competence (Berent, 1982). Consider Item A-3, which shows the child a picture of the numeral 7 and asks: "What numeral is this?" If the aim of the task is not obvious to a child, he may miss the item because he does not comprehend the term *numeral*. Children may not interpret Item A-9 in the same way as the testers intended and may be unjustly penalized as a result. The objective specifies: "Given a set of joined objects, count the total." A child is shown a picture of four blocks in the same orientation stacked on the bottom and three blocks in a different orientation stacked on top. Adjoining blocks are differently colored. The correct answer is seven. However, because blocks can be combined to form larger units, a child might interpret as a "block" the blocks sharing the same orientation and answer two.

The value of the KeyMath as a diagnostic instrument is limited because it focuses on the correctness of the child's answers (product) rather than how or why (the mental processes by which) the child got the answers. Even the finest level of analysis (item performance) is simply based on a child's right or wrong response to the item. The KeyMath does not have specific provisions for error analysis (Underhill et al., 1980). Though the manual suggests that the diagnostician determine the reasons underlying poor performance, it does not provide a theoretical framework for interpreting the results at any level of performance or for guiding follow-up procedures. In fact, the KeyMath is basically atheoretical in design. Thus, although the KeyMath Test may be useful for screening and classification purposes, it is not adequate as a diagnostic instrument (Underhill et al., 1980).

EDUCATIONAL IMPLICATIONS: COMPONENTS OF EFFECTIVE TESTING AND DIAGNOSIS

Testing based on absorption theory focuses on the number of items answered correctly; diagnosis, on which items are not answered correctly. To gauge understanding and thinking, cognitive theory points out that testing must go beyond scoring what pupils produce and examine process. Diagnosis should detect underlying difficulties, so that instruction can be adjusted, not merely repeated. Cognitive theory suggests that children's errors provide important clues about underlying processes and how to adjust remedial efforts.

For the most part, commercial and classroom standard tests do not provide the kind of diagnostic information necessary for planning remedial efforts. NRTs and other tests that focus on the number correct are useful in categorizing children as, say, high or low achievers, or as passing or failing. Diagnosis may begin by categorizing a child as failing, low-achieving, learning-disabled, and so forth. However, effective diagnosis goes beyond categorizing children and provides information that is useful to planning instruction (Glaser, 1981). To do so, diagnosis should indicate what a student has or has not learned and why learning has not occurred. CRTs and even standard tests that are called diagnostic tests are fine for initial diagnostic efforts only. To develop an effective remedial plan, a diagnostician must build and test a theory about a child's mental state. This necessitates more than just checking off what a child produces. Diagnosis involves generating hypotheses about a learning difficulty by, say, analyzing the child's work or probing his or her answers. After a remedial plan has been implemented, further evaluation serves to check the correctness of the hypothesis. In effect, diagnosis is an ongoing problem-solving process.

To build a theory about a child's mathematical knowledge as completely and accurately as possible, diagnostic efforts should collect data on informal knowledge, specific strengths and weaknesses, skill accuracy and efficiency, concepts, strategies, and errors.

1. *Effective diagnosis must examine informal as well as formal knowledge.* Children enter school with important individual differences in informal mathematical knowledge and hence readiness to learn formal mathematics (Baroody, 1983a; Baroody & Ginsburg, 1982b). Some children have had rich preschool mathematical experiences and can count to 100, can count a set of even 20 objects efficiently, and can mentally compare the magnitude of two numbers such as 32 and 33. On the other hand, some environmentally deprived, learning-disabled, and especially mentally retarded children come to school and—for various reasons— cannot even count to 10 or count small sets of 4 or 5 without error, let alone compare the magnitude of numbers like 6 and 7. Even within "homogeneous" mainstream or special education classes, there are often significant differences in informal knowledge.

It is especially important with children who are having difficulty learning school mathematics—children such as Adam—to evaluate informal knowledge. On the one hand, many children with low academic achievement may already know a good deal of informal mathematics that can be exploited to learn formal mathematics. By noting the informal knowledge children have already learned (and may take for granted), a teacher can help build their confidence in their ability to learn mathematics. On the other hand, some children—especially special education children—may have deficiencies in informal knowledge that lead to learning difficulties in school. It is essential to identify and remedy informal deficiencies quickly before beginning formal instruction and before the child is entangled in a spiral of failure.

2. *Effective diagnosis details a child's individual pattern of strengths and weaknesses.* With schooling, the range of individual differences in mathematical ability often becomes more and more pronounced. Some children learn number facts and arithmetic computational procedures and principles easily; others have to labor. Individual differences become especially apparent when place-value-related instruction is introduced. Knowledge of specific strengths and weaknesses is crucial for effective instructional planning, especially in the case of children such as Adam who are having learning difficulties. For example, an evaluation should specify that a child can read one- and two-digit numerals but not three-digit numerals—thus providing more specific direction for instruction or remediation. Knowledge of specific weaknesses permits an educator to individually tailor instruction, devising an effective IEP or selecting helpful remedial activities. Specific knowledge of strengths permits the educator to more effectively focus training efforts on what needs attention. Moreover, children's strengths can often be exploited to help remedy weaknesses.

In addition to providing clear direction for devising remedial procedures, a focus on the specific fosters a conviction that something can be done to correct a child's deficiencies (Baroody & Ginsburg, 1982b). By contrast, vague diagnoses such as "minimal brain damage," "learning disability," "perceptual dysfunction," or "mentally handicapped" do not suggest a clear or specific course of remedial treatment. In addition to being of little practical value, these general labels may be frightening or prejudicial, giving the impression that the child cannot be helped.

3. *Effective diagnosis evaluates accuracy and efficiency of skills.* Because school mathematics routinely builds on basic skills, it is important that a child use them accurately. Moreover, because many basic skills are often combined to form the integral parts of more complex skills, it is essential that such skills be automatic as well as accurate. For example, mastery of a multidigit multiplication algorithm requires efficient application of two component skills: recall of basic multiplication combinations and application of the written addition algorithm including renaming.

4. *A diagnostic test needs to assess concepts.* Because effective instruc-

tion and remediation involve fostering meaningful learning as well as the acquisition of skills, diagnosis should not end with evaluating skill accuracy and efficiency (Brownell, 1935). It is quite possible for children to get correct answers quickly but with little or no understanding of what they have done (cf. Brownell, 1935). For example, children may accurately imitate computational routines without understanding their place-value rationale. Unfortunately, many such children do not see how the formal procedures apply to new procedures or everyday problems. For this reason, some children forget all or part of the procedure.

5. *Diagnostic evaluations should examine solution strategies.* It is important to watch children's overt behavior and analyze their responses to gauge solution methods for a number of reasons (Ginsburg & Mathews, 1984). First, because children often do not do mathematics in the prescribed way, it may be helpful to know how a child informally copes with a mathematical task. This can provide insight into how the child understands the problems and what formal skills and concepts need to be taught. Second, examining strategy may indicate whether or not a child really understands a correctly used procedure or even uses a correct procedure to obtain correct answers. Third, examining a child's solution method may identify a specific unlearned step or misunderstanding that can be remedied efficiently. Fourth, examining process may reveal a performance failure that an incorrect test-item response does not indicate.

6. *Error analysis can be an important source of information about deficiencies in their underlying knowledge.* An analysis of systematic errors is an invaluable means of determining which step in an algorithm is causing a difficulty and what specifically needs to be retaught. This is especially important when new material is introduced. Quick diagnosis of procedural deficiencies can minimize or avoid practicing a procedure incorrectly and possibly establishing an erroneous habit pattern. Moreover, systematic errors may indicate an underlying misconception of misunderstanding that needs to be remedied. Thus an examination of systematic errors is critical to elaborating a theory about the internal state of a child and planning remedial efforts.

CHAPTER 5

Beliefs and Math Anxiety

How can a focus on memorizing facts and procedures contribute to the development of unreasonable and destructive beliefs? How can unreasonable beliefs affect a child's mathematical learning and problem solving? How can teachers foster more reasonable and constructive beliefs? How do beliefs contribute to math anxiety? How can math anxiety be treated?

THE NATURE OF THE PROBLEM

The Case of Mark

A woman pleaded desperately over the phone, "We're afraid that our 12-year-old son is learning-disabled—that there is something wrong with his brain. He's just unable to learn math. Would you examine him to see if there is anything that can be done for him?"

Shy about entering my office, Mark's first words revealed his fear: "So you're going to find out how dumb I am!" I explained that the purpose of the meeting was to find out what he did know about math and what he did not know so that his parents and teachers would be in a better position to help him learn. To help him feel more comfortable and to evaluate basic number fact and arithmetic skills, we then played several math games. Mark was perceptive: "These games—they're really tests, aren't they?" I explained the rationale for each test, pointed out that clearly he had learned much math already, and noted that doing math and having fun are not mutually exclusive. Satisfied, Mark began to relax and open up.

Mark outlined his math program and difficulties. The boy noted with despair that he could not remember the formulas for computing area. Questioning further revealed that he did not understand the concept of area.

To see if he was capable of learning this aspect of formal mathematics, the topic was introduced in a way that built upon his informal knowledge. I gave Mark a 4-by-3-inch cardboard rectangle, helped him mark off 1-inch lengths on the sides, and connected the marks to divide the rectangle into 1-inch squares. After Mark measured how big a square was (1 inch on each side or "1 square inch"), I asked him to figure out how many square inches the 4 × 3 figure had. Mark counted the squares and found 12. This process was repeated a number of times with rectangles of different dimensions. Asked if he noticed anything about how the dimensions of the sides and the area might be related, he thought for a while. Then suddenly he exclaimed with excitement, "You just have to multiply the two numbers!" I summarized the discovery: "You multiply the length and the width." (See Steps 1–4, Figure 5.1.)

Mark was then shown a cardboard parallelogram with a base of 4 inches and a height of 3 inches. Mark's enthusiasm for determining areas evaporated immediately. I empathized, "It looks impossible, doesn't it? But let's try looking at the problem in a more familiar way." I then cut off one of the protruding ends, placed it on the other end, and created a rectangle (see Step 5, Figure 5.1). Mark's response was immediate: "Oh, it's just base times height—12!" When topics were introduced in a meaningful manner (i.e., used an informal approach and built on previous knowledge), Mark appeared to be an alert and quick learner (Baroody, 1986).

In Mark's case, learning problems did not stem from a learning disability or brain dysfunction but from instruction that was not geared to his thinking. This instruction helped to create beliefs that undermined his desire and capacity to learn. He believed that mathematics was something children memorized—something that had little connection with his experience. Unlike some of his more successful classmates, Mark was not inclined to rotely memorize material that he did not understand or that he did not see as useful. Moreover, Mark believed he was incapable of understanding mathematics. Like many children, he blamed his failure on himself: "I can't do math because I'm stupid." Mark had given up on himself and even trying to learn school mathematics.

How had schooling contributed to Mark's belief that he was incapable of mathematics? Is it common for children to believe that mathematics is beyond their capabilities? What other debilitating beliefs does instruction that is not geared to children engender? What caused Mark to be so fearful of mathematics and the interview? How can children be helped to conquer debilitating fear?

The Learner as a Whole Person

Children come to school as complete human packages. Students are not merely cognitive (intellectual) machines that learn information. Their affective side (needs, drives, feelings, and interests) has a tremendous influence over their learning and use of mathematics (Reyes, 1984). Instruction

FIGURE 5.1: Meaningful Instruction of Area

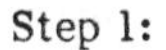

Step 1: Rectangle marked off in units of 1 square inch.

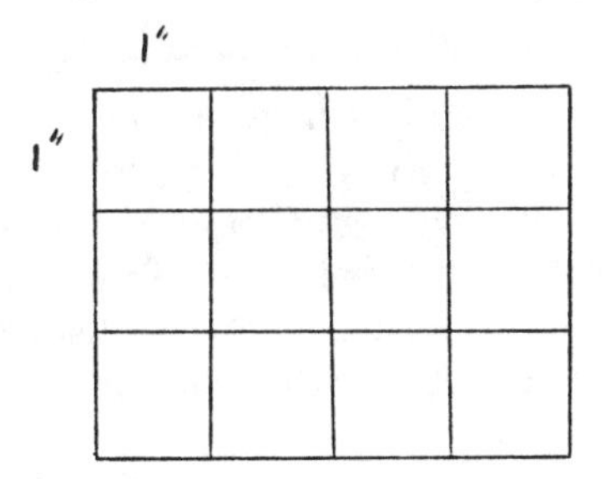

Step 2: Total number of 1"-square squares determined (by counting): 12.

Step 3: Discovery of Area = Length x Width encouraged by the suggestion to find a relationship between the informally computed area and the dimensions of the sides.

Step 4: Discovery formalized as a formula: A = L x W.

Step 5: The area of a parallelogram is reformulated in terms of a familiar problem.

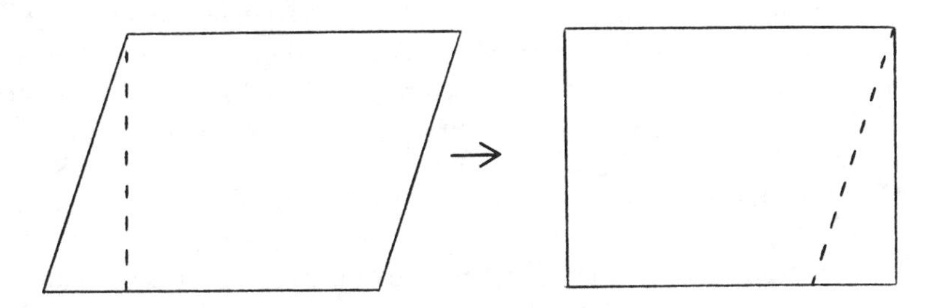

Step 6: Discovery of the connection between the area of a rectangle and parallelogram, formalized as the formula A = L x W.

Step 7: Repeat the process with different units (e.g., square centimeters) and different forms (e.g., squares and half a square or triangle).

that is not geared to the child may have harmful affective as well as intellectual consequences. Such instruction can dampen children's interest in mathematics. Indeed, it can so dishearten some children that they avoid and fail to learn mathematics altogether. For too many children—especially those labeled learning-disabled, slow learners, or low achievers—school mathematics seems to be beyond their grasp. This is further evidence of their inferiority and thus a threat to their very sense of well-being.

Beliefs are a key link between cognition and affect. Instruction that is not geared to children can distort their view of mathematics, mathematical learning, and self. These distorted beliefs can deaden children's drive to learn mathematics or create such fear that children cannot learn.

BELIEFS

Lessons Taught by an Absorption Approach

An overemphasis on memorizing facts and procedures in a lockstep manner cultivates debilitating beliefs. When instruction places a heavy emphasis on memorizing facts and skills, it is very likely that children will get the wrong impression of what mathematics is all about. If children are rushed into using abstract mathematical symbolism, they may conclude that understanding plays no role in mathematics (Hiebert, 1984). Instruction based on absorption theory teaches children "that learning mathematics is mostly memorizing" (Carpenter, Lindquist, Matthews, & Silver, 1983, p. 657) and "always gives a rule to follow to solve problems" (pp. 656–657). Cobb (1985a) points out that evidence (e.g., Baroody et al., 1983; Carpenter, Hiebert, & Moser, 1983; Cobb, 1985b; Lester, 1983) indicates that debilitating beliefs begin to develop in the primary grades.

The implicit or even explicit message of instruction based on absorption theory is: "Learn and use *the* correct facts and procedures and do so quickly." This direct or inferred message can foster perfectionistic beliefs:

- An inability to learn facts or procedures quickly is a sign of inferior intelligence and character.
- An inability to answer quickly or use a procedure efficiently indicates "slowness."
- An inability to answer correctly denotes a mental deficiency.
- An inability to answer at all signals real stupidity.

An overemphasis on getting *the* right answer using *the* correct procedure can create the following misconceptions as well:

- All problems must have a correct answer.
- There is only one (correct) way to solve a problem.
- Inexact answers (e.g., estimates) or procedures (e.g., trial-and-error problem solving) are undesirable.

Because instruction based on absorption theory too often overlooks, or worse, actively discourages children's informal mathematics, children learn to feel ashamed of their informal mathematics. When formal mathematics does not build on children's informal mathematics, they may feel that school math is beyond their comprehension and governed by its own inexplicable rules. Some beliefs engendered by not adequately considering children's informal mathematics are summarized below:

- Finger counting is childish and stupid.
- Mathematics is something only a genius can comprehend.
- Mathematics is not supposed to make sense.

A Cause of Mechanical Behavior

Beliefs promoted by an absorption approach discourage thinking and encourage blind procedure following (Holt, 1964). Such beliefs can interfere with meaningful learning and intelligent problem solving and cause children to go about learning and using school mathematics in a mechanical manner (e.g., Baroody, 1983a, 1986; Cobb, 1985b; Eccles, Adler, Futterman, Goff, Kaczala, Meece, & Midgley, 1983; Erlwanger, 1973; Fennema & Peterson, 1983; Schoenfeld, 1985; Tobias, 1978).

Unreasonable beliefs help explain why children fail to monitor their work and so readily accept inconsistent or nonsensical answers. Consider the common small-from-large bug. For a problem like 205 – 17 = __, many children answer 212! To answer correctly, the child would either have to know the borrowing algorithm or use an informal counting strategy. If a child does not know the borrowing algorithm and believes that the use of informal counting strategies is "bad" or "stupid," there are few options left. The child could make an estimate or not respond, but children frequently believe neither is acceptable. The option of last resort is to manufacture an answer by subtracting 5 from 7 and 0 from 1. This at least has the virtue of yielding some answer (cf. Holt, 1964). Thus, despite the fact it is inconsistent with their informal understanding that subtraction involves reducing an amount, they accept 205 – 17 = 212 without question. Because many children conclude that school mathematics is not connected with their informal knowledge and not supposed to make sense, they even fail to use what resources they do have available.

Unreasonable beliefs help account for why children do not respond thoughtfully to novel problems. For example, given a problem such as $3 - 7 = \_$, many primary-level children simply respond, "four." A correct response of " − 4" is not likely because most young children do not know about negative numbers. Children feel compelled to give the answer "four" because they have been trained to believe that there must be a correct answer and that giving some answer is better than saying, "I don't know" (Holt, 1964).

Children quickly learn to equate mathematics with arithmetic rather than searching for and defining relationships—intelligent problem solving. When given arithmetic problems, children believe they are supposed to calculate—*not* look for patterns, relationships, or shortcuts. As a result, many children may stop looking for regularities and fail to use what they know. Thus, despite their knowledge of commutativity, say, many children laboriously recalculate the sum of $18 + 27$ after just computing $27 + 18 = 45$.

Indeed, many children reach the conclusion that looking for shortcuts or doing anything but calculating is wrong or bad. Consider the results of a study (Baroody et al., 1983) that examined children's use of mathematical relations. The children were encouraged to solve sequences of problems in any way they wanted. The previous problem and recorded sum were left in a discard pile in open view. This information (e.g., $7 + 5 = 12$) was frequently useful in solving the next problem (e.g., a commuted version: $5 + 7$). Yet some primary-level children seem to feel that looking back at the previous problems to shortcut their computational effort was "naughty." For example, some furtively looked at the discard pile. A few looked and then provided a cover story (e.g., "I didn't look. I just remembered that one."). One girl explained, "I cheated on that one. I looked at the [previously computed sum]." Already these primary-level children had acquired beliefs that made them feel uncomfortable about intelligently searching out patterns and solving problems in an efficient but "nonstandard" manner.

MATH ANXIETY

Beliefs and Protective Behavior

Beliefs help explain why some children cope with mathematics while others become so anxious they become defensive. Consider a child who does not know the renaming algorithm but has reasonable beliefs: "It is all right not to know everything and to ask questions when help is needed." This child might conclude, "Oh, I don't know how to do this. I'll ask the teacher what needs to be done next." Now consider a child with unreasonable beliefs: "Smart people know everything; only dumb people have to ask questions." When this child runs into difficulty with the borrowing algorithm, the situation is interpreted very differently: "Oh, no, I can't do this. This must mean I'm dumb. I can't let anyone find out how dumb I am. I'll copy Sarah's answers—she always knows what to do." A child with unreasonable beliefs tends to become upset (anxious) and respond defensively (with protective behavior). Table 5.1 lists some of the protective strategies used as a result of the anxiety caused by three common unreasonable beliefs.

TABLE 5.1: Common Unreasonable Beliefs and Protective Strategies

Unreasonable Beliefs	Protective Strategies
1. Only dummies count to compute; only <u>really</u> dumb kids count on their fingers.	1A. <u>Covering up (count quickly and secretly)</u>: Decreases chance of being "discovered," but increases chances of error. 1B. <u>Guessing (or, as a last resort, not responding) to avoid counting solution</u>: Eliminates any chance of being "found out" but greatly increases chances of error.
2. Smart kids always answer correctly; it's dumb to answer incorrectly.	2A. <u>Not checking</u>: The illusion of perfection can be maintained, but imperfections are not discovered and corrected. 2B. <u>Procrastinating</u>: Real ability cannot be fairly evaluated if work is done in a last-minute rush, but quality of work suffers. 2C. <u>Covering up (e.g., mumbling)</u>: Act knowledgeable to maintain outward appearances, but needed assistance goes wanting. 2D. <u>Not trying</u>: The illusion of capability can be maintained because abilities are never tested, but failure is guaranteed.
3. Smart kids answer quickly.	3A. <u>Answering impulsively</u>: Throwing out the first answer that comes to mind helps maintain appearance, but is often wrong. 3B. <u>Covering up</u> (e.g., even when the answer is not known raise hand in response to teacher's question).

Children resort to protective strategies because of their short-term advantage. A protective strategy provides the immediate gain of minimizing anxiety. For example, by cheating a child can avoid the tremendous fear of being seen as inadequate. Unfortunately, the protective strategy will defeat the child in the long run. By cheating the child loses an opportunity to learn and become more adequate. Moreover, if the protective behavior is used for a long time, the child may be caught and made to feel even more inadequate. Table 5.1 briefly describes a short-term advantage and long-term disadvantage for each protective strategy.

A Model of Math Anxiety

Some children are so overwhelmed by fear that they become intellectually and emotionally paralyzed by mathematics. Math anxiety is a vicious cycle of unreasonable beliefs, anxiety, and protective behaviors. As the model in Figure 5.2 shows, anxiety begins with unreasonable beliefs (Ellis & Harper, 1975). These unreasonable beliefs lead children to overexagger-

FIGURE 5.2: A Model of Math Anxiety

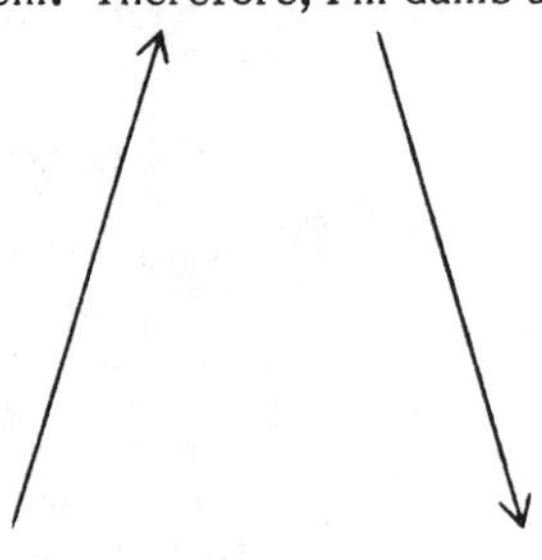

ate the importance of getting an answer and to underestimate their own self-worth. Because it threatens their whole sense of self, these children become terribly worried about their inability to solve any problem. The assignment of a mathematics problem becomes an anxiety-provoking situation. In fact, such children talk themselves into a panic (an anxiety attack). To contain the overpowering anxiety, a child resorts to a protective strategy, such as withdrawal from the anxiety-provoking situation. Not trying is self-protective to the extent that the child cannot get the problem wrong, which would prove his or her inadequacy ("Maybe if I had really tried I could have solved the problem"). On the other hand, by giving up, the child does not get the work done and receives yet another unsatisfactory grade—further evidence that the debilitating beliefs were correct. Thus the protective strategy makes facing the next assignment even more anxiety-provoking and increases the likelihood of the child's again responding in a self-protective manner. In brief, because of unreasonable beliefs, children can talk themselves into a self-sustaining, self-defeating, trap. Consider the following illustrative case.

The Case of Paul

An elementary school teacher was very upset about her third-grade son, Paul. In spite of efforts to help him, Paul was over a year behind in mathematics achievement, and the situation was growing worse. Reluctance to doing mathematics had become resistance to doing mathematics. In addition, Paul had begun to break out in hives. These psychological and physical symptoms stemmed from the stress caused by math anxiety.

A key to Paul's problem became evident in our first session. Paul was extremely sensitive about calculating with his fingers. To calculate sums, differences, and products, Paul held his fingers as still as possible. Moreover, he held his hands under the table out of sight. Criticism of his informal arithmetic at school and at home had made Paul anxious and wary about using his fingers. Paul had come to believe that informal arithmetic was a crutch that only the intellectually incapable used. In effect, he equated finger counting with stupidity.

Because of the tremendous anxiety it caused, Paul preferred to avoid finger counting. For Paul, this essentially meant avoiding school mathematics altogether. Though inaction minimized anxiety about finger counting in the short term, it only made the situation worse in the long run: It guaranteed lack of progress and failure. Lack of progress and failure confirmed his fears that he was intellectually incapable. This made him even more anxious about finger counting and tackling the next assignment, and so he resisted trying all the more.

EDUCATIONAL IMPLICATIONS: FOSTERING CONSTRUCTIVE BELIEFS

"The medium is the message" (McLuhan, 1964, p. 7). *How* mathematics is taught says as much or more about mathematics as *what* is taught. Young children are prone to draw overdefinite conclusions, especially when presented with a one-sided case. Young children are prone to accept uncritically the demands or judgments placed upon them by authority figures. An overemphasis on efficient memorization and use of facts and skills, then, can create a distorted view of mathematics, mathematical competence, and self. It can foster misconceptions, perfectionistic beliefs, and a sense of inadequacy that can undermine children's intelligent learning and use of mathematics. Instruction based on absorption theory tends to cultivate beliefs that promote blind procedure following over thinking, mechanical behavior over thoughtful monitoring and problem solving. Clearly, it is important to teach children "the basics," but this should not be done in a way that encourages debilitating beliefs—beliefs that compromise the development and use of thinking.

For the sake of meaningful learning, intelligent use of mathematics, and the well-being of students, it is essential to encourage reasonable and constructive beliefs about mathematics, mathematical learning, and self. It is possible to encourage healthy beliefs in children who have already begun to learn unhealthy ones. Indeed, it is possible to teach even math-anxious children more constructive beliefs and thus break the vicious cycle of self-defeating behavior. In fact, it may be necessary to change some of the entrenched beliefs of the math-anxious child before an effort is made to remedy skill deficiencies. Specific guidelines and examples are described below.

1. *Point out the inaccuracy of perfectionistic beliefs and help children develop perspective.* Avoid overemphasizing perfection, particularly the need always to get the right answer. Holt (1964) tried to teach constructive beliefs by using analogies. For instance, a good baseball batting average is .333, which means the player got a hit only one in three times at bat. The point is that even capable people are not perfect. Another analogy compared not asking questions about a puzzling topic with leaving something at Howard Johnson's on a long trip: You'll have to go back for it, and it will be easier if done as soon as possible. The moral is that in the long run, it is more intelligent to ask questions when puzzled than to maintain an appearance of understanding.

Explicitly and implicitly counter common misconceptions. It is important to counter common debilitating beliefs by both word and deed. For example, because using relationships to shortcut computation is so often viewed as cheating, explicitly encourage children to look for and use mathematical regularities (Baroody et al., 1983). Moreover, these words need to be backed with actions. Children should be praised for finding and sharing shortcuts, exercises should be given with the expressed aim of finding and exploiting relationships, and so forth.

To counter the misconception that all problems must have an answer, a teacher might point out that it is not always possible to answer mathematical problems. For example, scientists can make an educated guess about how many stars the universe contains, but an exact answer is beyond our means to count and is always changing anyway (stars are continuously forming and dying). Children need exercises that regularly require "not possible" as a response. For example, in the Wynroth (1969/1980) curriculum, worksheet exercises on finding the multiplication factors (other than 1 and *N*) of numbers would include both solvable problems (e.g., 24: factors other than 1 and 24 are 2, 12; 3, 8; and 4, 6) and nonsolvable problems (e.g., 29, for which there are no factors other than 1 and 29). In effect, the curriculum deliberately teaches children that not all mathematical problems are solvable—an important belief in developing the proper perspective on mathematical problem solving.

To promote the idea "that mathematics involves understanding and gaining insights rather than finding ways to give the impression" of knowing the correct answer or procedure (Cobb, 1985a, p. 144), a teacher should encourage children to talk with each other and the teacher about mathematics. Small-group problem-solving exercises (e.g., Easley, 1983) can also nurture the idea that mathematics involves thinking (Cobb, 1985a).

2. *Relate new material to experiences familiar to children.* Divorced from informal mathematics, formal mathematics often appears foreign—an overwhelming obstacle that children feel their meager resources cannot overcome. In other words, many children feel alienated from school mathematics because they have been robbed of their sense of control over it. Relating school mathematics to informal mathematics should make it seem less strange, threatening, and overwhelming, and may help children to feel in charge of their mathematical learning. This greater sense of control could result in children's feeling better about mathematics, school, and themselves.

3. *Encourage a positive view of children's informal mathematics.* This is especially important in the case of children with learning difficulties or math anxiety (Baroody, 1986). Children are often made to feel ashamed of

their informal strategies by teachers, parents, peers, siblings, or others. As a result they try to hide or disguise their informal strategies. Worse, they begin to believe that their informal approaches are invalid—that their way of thinking about mathematics is inadequate and stupid. These problems can be avoided, or at least minimized, if the following suggestions are incorporated into the teacher's approach:

- *Present an accepting attitude toward informal mathematics.* This action by itself removes a tremendous barrier between children and teacher and is a crucial first step in helping children with math difficulties. In effect, the teacher tells the child, "Your approach to math—your thinking—is *not* stupid and hopelessly deficient." By building this bridge, the child in trouble has a basis for trusting the teacher. As a result, it is much more likely that the child will be more responsive to instruction.
- *Note informal strengths and the value of this informal knowldge.* Children, like most adults, take their informal knowledge for granted, without realizing how sophisticated and powerful it is. It is sometimes helpful to compare a child's informal knowledge with that of a younger child. For example, give the child a card with 10 dots and one with 11 and ask the child to figure out which card has more. Encourage the child to count, if need be. Then point out that younger children do not know that 11 is more than 10, and that very young children cannot even count a set of 10 or 11 dots correctly!
- *Help develop perspective about informal mathematics.* A teacher can attack unconstructive beliefs about informal mathematics directly by pointing out the important role informal mathematics played in the historical development of mathematics. It is quite a revelation for children (and most adults) to learn just how important, say, finger counting has been in the development of mathematical knowledge throughout human history (e.g., see Chapter 2). Such discussions can be reassuring to children—especially older children having difficulty or math-anxious children who rely on informal arithmetic.

Consider the case of Paul, described earlier. One of the first objectives of remedial efforts was to change Paul's beliefs about informal arithmetic by encouraging rather than criticizing his informal arithmetic, pointing out the power of his informal knowledge, and describing the central role of counting in the historical development of mathematics. Paul rather quickly became more open and relaxed. Probably as important, Paul's mother and teacher became more accepting of his informal strengths. Several months later, Paul's mother reported that both she and Paul were more comfortable with his math progress. Paul had stopped breaking out in hives, and his new achievement scores were much closer to grade level. By substituting constructive beliefs (about informal mathematics in particular) for unreasonable beliefs, the vicious cycle of math anxiety was broken. Reasonable beliefs gave Paul perspective and did not stir up undue anxiety. Less anxious about his informal mathematics, he had less need to resort to protective strategies. Moreover, freed from needless worrying, Paul applied himself to the task of learning.

Part II

INFORMAL MATHEMATICS

CHAPTER 6

Counting Skills

Does oral counting imply numbering skill? What counting skills typically develop during the preschool years? Can we assume that special education children will acquire basic counting skills informally? What skills typically require training in the first years of school?

THE DEVELOPMENT OF COUNTING SKILLS

The Case of Alexi

By 26 months, Alexi could orally count from 1 to 10 and had begun experimenting with the teens. Asked to count three dots in a triangular formation, Alexi pointed at the dots and rattled off: "1, 2, 3, 4, 5, 6, 7, 8, 9, 10." Asked to count three dots in a row, he randomly pointed at the set several times as he uttered: "8, 9, 10." Even after he could accurately count sets of up to five objects, Alexi was puzzled when asked how many he had counted. When shown two sets (e.g., a card with nine dots and one with eight dots), he was also baffled by requests to point to the one with "more."

Alexi's oral-counting skill did not guarantee an ability to accurately count sets of objects or other number skills. Yet by 5 years of age,[1] children can not only orally count to about 29 but immediately determine that • ˙ • and • • • are "three." Moreover, it is obvious to a typical 5-year-old *how* to solve the problem of determining which of two sets (e.g., nine vs. eight dots) has more: Just count each set and compare the resulting numbers. After counting each set of dots, the *solution* to the problem is also readily apparent to the 5-year-old: "The set with 9 is more." Thus in a matter of a few years, children learn a range of counting skills and much about how to

1. The behaviors described below are based on the norms of the Test of Early Mathematical Ability (Ginsburg & Baroody, 1983) and represent the "average" ability of a child 4 years and 11 months old.

use counting (Fuson & Hall, 1983). Just how complicated this development is—just how much adults come to take for granted—is revealed by a close examination of the skills mentioned in the paragraph above.

A Hierarchy of Skills

For the most part, counting ability develops in a hierarchical fashion (Klahr & Wallace, 1973). With practice, a counting skill becomes more automatic and requires less attention to execute. Once a skill can be executed efficiently, it can be simultaneously processed or integrated with other skills in working (short-term) memory to form an even more complex skill (e.g., Schaeffer, Eggleston, & Scott, 1974). Consider what is required by the apparently simple task of determining whether a set of nine dots or eight dots is more. To make such a numerical magnitude comparison involves the integration of four skills.

First, the most basic skill is consistently generating the number words in their proper sequence. Two-year-old Alexi had already begun to master the *oral number sequence* and could sometimes count by ones to 10. However, when asked to count objects, he did not yet consistently say the numbers in the correct sequence. For example, he sometimes did not begin his counts with "one." By three years or so, children typically begin counting a set with "one," and by the time they enter kindergarten, they can use the correct sequence to count sets up to at least 10 (Fuson, Richards, & Briars, 1982).

Second, the number-sequence words (tags) then have to be applied one at a time to each object in a set. Object counting is called *enumeration*. Though he could generate the correct number sequences to 10, Alexi could not enumerate a set of nine or even three because he had not yet learned that one and only one tag should be applied to each item of a set. Enumeration is an involved skill because the child must coordinate saying the number sequence with pointing to each item of a collection so as to create a one-to-one correspondence between tags and objects. Because 5-year-olds can efficiently generate the number sequence and point once to each item in a collection, they can effectively coordinate the two skills to execute the complex act of enumeration (at least with sets up to 10).

Third, to make a comparison, a child needs a convenient way to represent how many dots each set contains. This is accomplished by the *cardinality rule*: The last number tag uttered in the enumeration process stands for the total number of items in the set. In other words, a 5-year-old can simply summarize the count chain "1, 2, 3, . . . 9" with "nine" and the count chain "1, 2, 3, . . . , 8" with "eight." Because Alexi could not even enumerate sets, he had not discovered that the last tag in this process has special significance. For 2-year-old Alexi, the number sequence was not yet associated with defining the quantity of a set.

Fourth, the three previously described skills are a prerequisite for an understanding that position in the sequence defines *magnitude*. For 2-year-old Alexi, numbers did not define relative size. However, young children eventually learn that the sequence of numbers is associated with relative magnitude. Even very young children can make accurate gross magnitude comparisons such as "10 is bigger than 1," perhaps because they know that 10 comes much later in their count sequence. By 5 years, children can even quickly make fine magnitude comparisons between "number neighbors," such as nine and eight, because they have great familiarity with number-after relationships ("nine comes after eight when I count, so nine is bigger").

Thus, counting to determine that a set of nine dots is more than a set of eight dots is not, cognitively speaking, a trivial act. Though adults may take for granted the four skills involved, they are an imposing intellectual challenge to the 2-year-old. By the time they are five, most children will have mastered these basic skills and be ready for new challenges. Some—notably environmentally deprived, brain-injured, or mentally retarded children—may not have mastered these basic skills and will need special help. The rest of the chapter describes in more detail the four basic counting skills and more elaborate counting skills that develop in the first years of schooling.

Oral Counting

NUMBER SEQUENCE. As early as 18 months, children begin to count orally by ones ("one, two, three . . . "). The majority of two-year-olds can count "one, two" but then begin to omit terms (Fuson et al., 1982). Initially, children may learn chunks of the sequence to 10, which are then strung together. For example, Alexi (about 20 months) first used—on a regular basis—the chunk "8, 9, 10." Later he added the chunk "2, 3, 4," to make "2, 3, 4, 8, 9, 10." Later still, he added 5 and 6 and finally 1 and 7 to fill in the number sequence to 10. At 26 months, Alexi added the two-digit terms 19 and 20 and, soon afterward, he inserted the chunk "11, 12, 13" between 10 and 19.

Oral counting is often called "rote counting." As the case of Alexi illustrates, rote counting is an apt description of children's earliest oral counting. His oral counting was simply a meaningless verbal chant. Alexi's initial number sequence appeared to be nothing more than a chain of associations that were rotely memorized and gradually linked together. However, rote counting is a less apt description of later counting efforts. The term is too often taken to mean that children learn the whole number sequence by rote memorization. Though rote memorization plays a role, especially in the initial stages of learning, rule-governed learning is central to extending the sequence. Though the terms up to about 13 are probably rotely memorized, the number sequence thereafter can, for the most part, be generated by rules (Ginsburg, 1982). Except for fifteen, the remaining teens can be generated by continuing with the original sequence (4, 6, 7, 8, 9,) and adding teen (e.g., "six-teen, seven-teen, . . . "). The twenties (21, 22, 23, . . . , 29) can be generated by the rule: combine twenty with each of the single terms (1 to 9) in turn. Indeed, a child has only to learn this rule and the decades (10, 20, 30, . . . , 90) in order to count by ones to 99.

Children's counting errors are a good indication that rules underlie oral counting, especially for 20 and above. Many children—including mentally retarded children—substitute made-up terms such as "five-teen" for fifteen, "ten-teen" for twenty, or "twenty-ten, twenty-eleven" for thirty, thirty-one (Baroody & Ginsburg, 1984; Baroody & Snyder, 1983; Ginsburg, 1982b). Such errors clearly indicate that children are not merely imitating adults but are struggling to construct their own system of rules (Baroody & Ginsburg, 1982). These are sensible errors in that they are logical, though incorrect, extensions of the number-sequence patterns the child has abstracted. Thus, even mentally retarded children appear capable of seeing, using, and sometimes misapplying the patterns in the number sequence.

Though most children just beginning school are already making progress with the rule-governed portion of the number sequence, many do not realize that the decades ("10, 20, 30, . . .") follow a pattern paralleling the one-digit sequence (Fuson et al., 1982). How children solve the "decade problem"—the correct order of the decades so as to count to 100 by ones—remains an open question. One hypothesis is that children learn the decades rotely as the end items for each series (e.g., the child rotely forms the association of "29-30" or "39-40"). There is some support for this conjecture. Some children cannot count by tens but are able to count to 30 or 39, because they have apparently learned that 30 follows 29 but have not learned what follows 39 (Baroody & Ginsburg, 1984). A second hypothesis is that children learn the decades (to count by tens) by rote and use this knowledge to fill in the count by ones sequence. A third and dramatically different hypothesis is that children learn the decades as a modified version of the original 1 to 9 sequence and use this pattern (repeat the one-digit sequence and add *-ty*) to fill in the ones count. This third hypothesis was illustrated by Teri, a mildly retarded girl, who would get to the end of a series (e.g., ". . . 58, 59") and then count to herself to figure out the next decade (e.g, "1, 2, 3, 4, 5, 6—ah . . . sixty") (Baroody & Ginsburg, 1984). This procedure was repeated until she got to 100.

In fact, most children may rotely memorize some of the decades (Hypothesis 1 or 2) and use rules to generate the rest (Hypothesis 3). This makes sense because most of the decades follow a pattern, so it would be inefficient to rotely memorize all of them. However, children may have to rotely memorize the first portion, including perhaps some regular cases such as forty, before discovering the pattern. Thus, learning to produce the decades (counting by tens) may be like learning to count by ones: Children first acquire a rote portion and then use a pattern to extend the count.

ELABORATIONS ON THE NUMBER SEQUENCE. With experience, children learn to use their mental representation of the number sequence in increasingly elaborate and flexible ways (Fuson et al., 1982). As the correct sequence of numbers becomes more familiar, children can specify number successors. That is, they automatically cite the number just after a given number. At 26 months, Alison could already give the number just after *if* she were given a "running start."

FATHER: Alison, what comes after nine?
ALISON: [No response.]
MOTHER: 1, 2, 3, 4, 5, 6, 7, 8, 9 . . .
ALISON: 10.

Otherwise, Alison was either unsuccessful or only occasionally correct.

MOTHER: What comes after eight?
ALISON: Eight.
MOTHER: What comes after two?
ALISON: Nine.
MOTHER: What comes after six?
ALISON: [No response.]
MOTHER (Somewhat later): What comes after eight?
ALISON: 9, 10.
MOTHER: What comes after two?
ALISON: Four.

By 4 or 5 years, children no longer have to start with one to answer number-after questions consistently and automatically—at least up to about 28 (Fuson et al., 1982; Ginsburg & Baroody, 1983). A development that may occur somewhat later is the ability to cite the numbers just before. Once children grasp number-just-before relationships, the stage is set for counting backwards. Moreover, school-age children gradually learn various skip counts. Among the earliest of these new patterns are counting by twos, fives, and tens.

Numbering

ENUMERATION. Children have to learn that object counting involves more than just shaking their finger at a set or sliding their finger over a set while they rapidly utter the number sequence. Though young children quickly learn at least the rote portion of the number sequence (see, e.g., Fuson & Hall, 1983) and have no problem in pointing to objects one at a time (Beckwith & Restle, 1966), coordinating these two skills to enumerate a set is not a trivial task. Indeed, enumeration, especially of sets larger than four, is a competence that only gradually becomes automatic (Beckwith & Restle, 1966; Gelman & Gallistel, 1978; Schaeffer et al., 1974). With large, disorderly collections in particular, children have to learn strategies for keeping track of tagged and untagged items. When items are aligned in a row, little effort is required to keep track if a child starts at one end of the row. If a collection forms a circle, a child need only remember with which item the count began. With disorderly arrays, a child must remember which items have been tagged and which need to be tagged. This is facilitated by using a systematic approach (e.g., moving from left to right, top to bottom) or by separating tagged items from untagged items. Fuson (in press) found that many of her kindergarten-age subjects did not use the strategy of creating a separate "counted pile."

CARDINALITY RULE. Initially, children may not realize that enumeration serves the purpose of numbering. When asked to count a set, children just enumerate the set and expect that this, in itself, will satisfy the adult (and it sometimes does). If asked how many objects they counted, they just reenumerate the set. For example, 3-year-old Ida enumerated four stars ("1, 2, 3, 4") without any real intent to use or remember the information. When asked how many stars she had counted, Ida shrugged her shoulders and enumerated the stars again. Because enumeration is viewed as an end in itself and not as a means to an end, very young children may not understand the point of how-many questions or bother to remember their counts.

At approximately 2 years, many children develop a primitive awareness that counting is a procedure used to assign numbers to collections (to answer how-many questions). They now make the effort to remember the count. However, because they do not realize that the enumeration process can be summarized, they respond to how-many questions by reiterating the number sequence. After spewing various terms ("7, 8, 9") or repeating the same term ("9, 9, 9") for, say, a set of three objects, a 2-year-old may designate the sets by repeating the count (e.g., "7, 8, 9" or "9, 9, 9") (Wagner & Walters, 1982). Even after they learn to enumerate accurately, children may not realize that it is unnecessary to recite the whole sequence when asked about how many. For example, after enumerating four stars pasted on a card, George (without looking at the card again) responded to the how-many question with: "Oh, there are 1, 2, 3, 4 stars." As early as 2½ years, though, some children discover the shortcut of just reciting the last tag in the enumeration process to indicate how many. In effect, the cardinality rule translates a count term that is applied to a particular (the last) item in a set into a cardinal term that represents the whole set.

CARDINAL-COUNT RULE. The reverse of this cardinality rule is the cardinal-count rule. The cardinal-count rule specifies that a cardinal term, say, "five," would be the count tag assigned to the last item when enumerating a set of five objects (Fuson & Hall, 1983). It seems that children have to learn that a term like *five* is simultaneoulsy a name for a set (cardinal number) and a count number. Consider the situation where a child is given a set of, say, five marbles, and instructed, "Here are five marbles; put five marbles in the cup." A child who does not appreciate the cardinal-count rule actually has to count the marbles as he puts them in the cup. This child cannot anticipate that the label *five* that designated the set would be the same as the outcome of counting the set. In contrast, a child who takes the cardinal-count rule for granted will—without counting—simply put the set into the cup.

PRODUCTION. Counting out (producing) a specified number of objects is a skill we use every day (e.g., "Get me three pencils," "I'll take four shirts," "Take five nails."). Yet it is not a cognitively uncomplicated task because it involves (a) noting and remembering the requested number of items (the target); (b) labeling each item taken with a number tag; and (c) monitoring and stopping the counting-out process. In other words, production requires storing the target in working memory, an enumeration process, while simultaneously checking the numbers in the enumeration process against the target number and stopping the count process when a match is achieved (Resnick & Ford, 1981). The cardinal-count rule gives the child reason to register the target in working memory and provides the basis for stopping the enumeration process (Baroody & Mason, 1984). For example, if a child is asked to count out three crayons, she must realize that it is important for the successful completion of the task to remember "three" and that she must stop counting crayons when she gets to the tag "three."

Magnitude Comparisons

At about 3 years, children discover that higher count terms are associated with larger magnitudes (Wagner & Walters, 1982). Thus, they realize that "two" not only follows "one" in the number sequence but also represents a larger quantity than does "one." On average, by $3^1/_2$, children appreciate that "three" is greater than "two" (Schaeffer et al., 1974). With these facts, children about age 4 seem to discover a general rule: The number term coming later in the sequence signifies "more" than a preceding number term does. Even before they enter school, children seem to use their mental representation of the number sequence to make efficient gross-magnitude comparisons—to compare quickly and accurately two numbers that are rather far apart in the sequence (e.g., 3 vs. 9 or 2 vs. 8) (Resnick, 1983). As "just-after" relationships become automatic, children may become capable of making fine-magnitude (number-neighbor) comparisons. Indeed, by the time they enter kindergarten, most children are quite accurate in making number-neighbor comparisons up to 5 and even 10.

EDUCATIONAL IMPLICATIONS: COUNTING DIFFICULTIES AND REMEDIES

Oral Counting

NUMBER SEQUENCE. Most children—including minority and lower-class children—receive extensive exposure to the first or rote portion of the number sequence from family, friends, nursery-school teachers, television, and so forth before entering school. The inability of a beginning kindergartner to generate the rote sequence to at least ten may indicate a severe problem and a need for immediate and intensive remedial efforts (Baroody & Ginsburg, 1982b). Though there are wide individual differences, mastery of the rote-counting portion should not be taken for granted in elementary-age retarded children (Baroody & Ginsburg, 1984). Most $4^1/_2$- to 6-year olds can even count to about 29 or 39. However, because they have not solved the decade problem, many are unable to extend their rule-governed portion beyond this point. Many young mentally handicapped children will need help mastering even the first portion of the rule-governed sequence (the teens and the twenties).

Beyond 13 or so, number-sequence training should not emphasize rote memorization. Instead, children should be encouraged to look for and discuss the patterns underlying the number sequence. In some cases, the teacher may have to provide hints or help make the patterns explicit (see Example 6.1). Furthermore, teachers should take heart when a child makes a rule-governed error such as substituting "twenty-ten" for thirty. This is an encouraging sign, for it suggests recognition of a number pattern and an active attempt by children to deal with the unfamiliar in terms of their existing rules or understanding. When a child makes a rule-governed error, the teacher can build upon the child's existing knowledge by saying, for instance, "*Another name* for twenty-ten is thirty." This is a constructive way to correct the child because the teacher shows appreciation for the child's thinking ability and yet provides the feedback needed for further growth.

The most common stumbling blocks for children of all mental abilities are the irregular teen *fif*-teen and the irregular decades *thir*-ty and *fif*-ty

EXAMPLE 6.1

Using Patterns to Teach the Decades

Even moderately retarded children may benefit from instruction that exploits the patterns underlying the number sequence. Consider the case of Mike, a 20-year-old with an IQ of 40. Mike was trying to learn how to tell time to the nearest 5 minutes. However, because he did not know the decades beyond 30, he could not count by fives past 35. After 35, he simply repeated previously used terms (e.g., "5, 10, 15, 20, 25, 30, 35, 30"). To make a connection between the single-digit sequence and the decades, Mike's tutor wrote down the number sequence 1 to 6 on a card. Underneath each single-digit term, she wrote the corresponding decade. Then she explained that he could use the first counting numbers to figure out the decades. "See, one is like ten, two is like twenty, three is like thirty, four is like forty, five is like fifty, and six is like sixty." Mike was able to use the card with the corresponding number lists to count by "fives." By using the card, he could count around the clock. Mike was so pleased with the method that he asked for copies of the card to use in his classroom and at home. The next steps were to encourage Mike to use his mental count sequence to figure out the next decade and then practice counting by tens and fives until these skills were automatic. At this point, Mike would be ready for telling time without the aid of the card.

Cathy A. Mason tutored this child and compiled the case study.

(e.g., Baroody & Snyder, 1983; Fuson et al., 1982). Because 15 is an exception to the teen pattern, it is often the last teen term learned. Some children will simply skip 15 (" . . . 14, 16 . . . ") or make a substitution (" . . . 14, 14, 16 . . . "). Expeditious diagnosis, modeling, and practice can habituate the proper sequence before an incomplete or incorrect sequence becomes established.

ELABORATIONS ON THE NUMBER SEQUENCE. By the time they are in kindergarten, children should have no problems citing the number just after or even the number just before a given number, up to at least 10 (Fuson et al., 1982; Ginsburg & Baroody, 1983). Low-functioning and mentally retarded children may not be able to cite the number just after and may have to count from one or resort to guessing. Citing the number just before is likely to be relatively difficult because the child must operate on the number sequence in the opposite direction from which the sequence was learned. Furthermore, *before* may be a more difficult term to understand than *after*. Thus it might be best to first focus remedial efforts on number-just-after. Remedial efforts should begin with the most familiar portion of the number sequence (one to four or five). Moreover, if the child can read numerals, it may help to start with activities that involve a concrete representation of the number sequence (a number list). After the child understands the after (before) question and can easily produce answers with the number list, shift to using activities without a number list that require the child to mentally determine the answer.

Counting backward from 10 depends on knowledge of number-just-before relationships and is a relatively difficult oral-counting skill. Nevertheless, it is one that is typically mastered by children by the time they enter first grade (Fuson et al., 1982; Ginsburg & Baroody, 1983). Counting backward from 20 is an especially difficult skill and is typically not mastered much before the beginning of third grade. Teachers of special education children can expect much difficulty with both skills. Remedial efforts can begin by having a child read a number list backward (from right to left). With children who are mastering or have mastered number just after, cover the number list except for the starting point. Then, as a child counts backward, reveal successively smaller numbers. This procedure affirms correct responses and provides corrective feedback for incorrect responses.

With skip counts of up to five at least, children can be encouraged to use their familiar count-by-ones sequence, but whisper intervening terms and emphasize the terms of the pattern. For example, to learn the twos, the child could be encouraged to count: "one [softly], *two* [loudly], three [softly], *four* [loudly]. . . . " If need be, this can first be done with a number list, which would free the child from the effort of producing the correct term and allow him or her to focus on the pattern. Example 6.2 illustrates another method for introducing skip counting in terms of the familiar count-by-ones sequence.

Numbering

ENUMERATION. By kindergarten age, children are generally quite proficient in counting sets of one to five objects, and most 5-year-olds will accurately enumerate up to 20 objects (Fuson, in press). Thus, a beginning kindergartner who is still having difficulty with sets of one to five needs immediate, individual attention. A child who makes no attempt, even with smaller sets, to label each object with one counting word (just spouts counting words as a finger glides over objects) or keep track of counted and uncounted objects (unsystematically tags set objects) is at considerable risk (Baroody & Ginsburg, 1982b).

Because enumeration requires the coordination of two subskills, errors may arise from three sources: (a) generating an incorrect number sequence

EXAMPLE 6.2

Skip-Counting Instruction

Skip counting can be made meaningful for children by relating it to the familiar procedure of counting real objects by ones. Josh, a moderately retarded adolescent, was learning to count by fives. His tutor had him put small, colored plastic disks he liked in piles of five and then helped him count the piles by fives. Next, she had Josh spread out the disks and count by ones. Josh was amazed that he got the same result. He then tested the generality of this finding with different numbers of piles. In his next tutoring session, Josh insisted on repeating the experiment himself.

In the third session, Josh asked for number cards (5, 10, 15, 20, 25, etc.) and matched these to his piles. He now added a new step to his checking process: reading the number cards as the disks were counted by ones. He checked the outcome of counting the first pile by ones against the first number card and found both were "5." Continuing the count by ones with the second pile, he found that this outcome matched the second number card: 10, and so forth. As Josh counted by ones, the tutor emphasized each of the row-ending numbers (5, 10, 15, etc.) by saying these with him. Josh then made up a guessing game in which he hid his eyes, the tutor took one card (e.g., 15), and he had to figure out which one it was. By the fourth session, he was able to count to 30 by fives independently. Use of real objects and his count-by-ones sequence made skip counting both understandable and interesting to Josh.

Cathy A. Mason tutored this child and compiled the case study.

(*sequence errors*), (b) inaccurately keeping track of counted and non-counted items (*partitioning errors*), and (c) not coordinating the production of the number sequence and the keeping-track process (*coordination errors*) (Gelman & Gallistel, 1978). Examples of each type of error are illustrated in Figure 6.1. Children may occasionally make a slip in generating numbers, but a consistent sequence error (e.g., regularly labeling sets of both 14 and 15 as "14") indicates that remedial efforts need to be directed toward the prerequisite oral-counting skill. A child who regularly makes partitioning errors such as skips and double counts (twice points at and tags an item) needs to learn more efficient keeping-track strategies.

In Figure 6.1, note that very different types of errors may produce the same answer. For example, like double counting, double tagging (pointing once to an object and letting two tags slip out) overvalues a set by one. However, double tagging is a coordination, not a partitioning, error. Indeed, several errors may combine to produce the correct answer. Because wrong answers can occur in various ways and because mathematically two wrongs do not make a right, it is important for teachers to observe the enumeration activity of pupils who are having difficulty.

If a child has trouble efficiently executing either subskill, coordination errors are likely. For example, a child who has to stop and think what comes after three when counting a set of five pennies may lose track of his place: "1 [tags first item], 2 [tags second item], 3 [tags third item], ah, ah, ah, yeah, 4 [skips fourth item and tags fifth item]." Likewise, if a child has to focus a great deal of attention on keeping track, the child may misproduce the number sequence (e.g., skipping a number). Fuson and Mierkiewicz (1980) found that young children were prone to make coordination errors in the middle of their counts.

Coordination errors can also occur at the beginning or the end of the enumeration process (Gelman & Gallistel, 1978). Some children have difficulty getting both subskills started at the same time. As a result, they point at but do not tag the first item or they start the tagging process too soon (e.g., say "one" without pointing to the first item, which is then tagged "two"). Sometimes children have difficulty ending the two subskills in coordination and thus point at but do not tag the last item or continue the tagging process even after the last item has been pointed at. Mentally retarded children appear prone to make coordination errors (Baroody & Ginsburg, 1984).

Two severe enumeration errors are "flurries," where the child may start the one-to-one process but then not maintain it, and "skims," where no effort is made to execute a one-to-one process when beginning or finishing the enumeration process (Fuson & Hall, 1983). Flurries can entail a failure to keep track of tagged and untagged items (a partitioning error), coordi-

FIGURE 6.1: Examples of Enumeration Errors

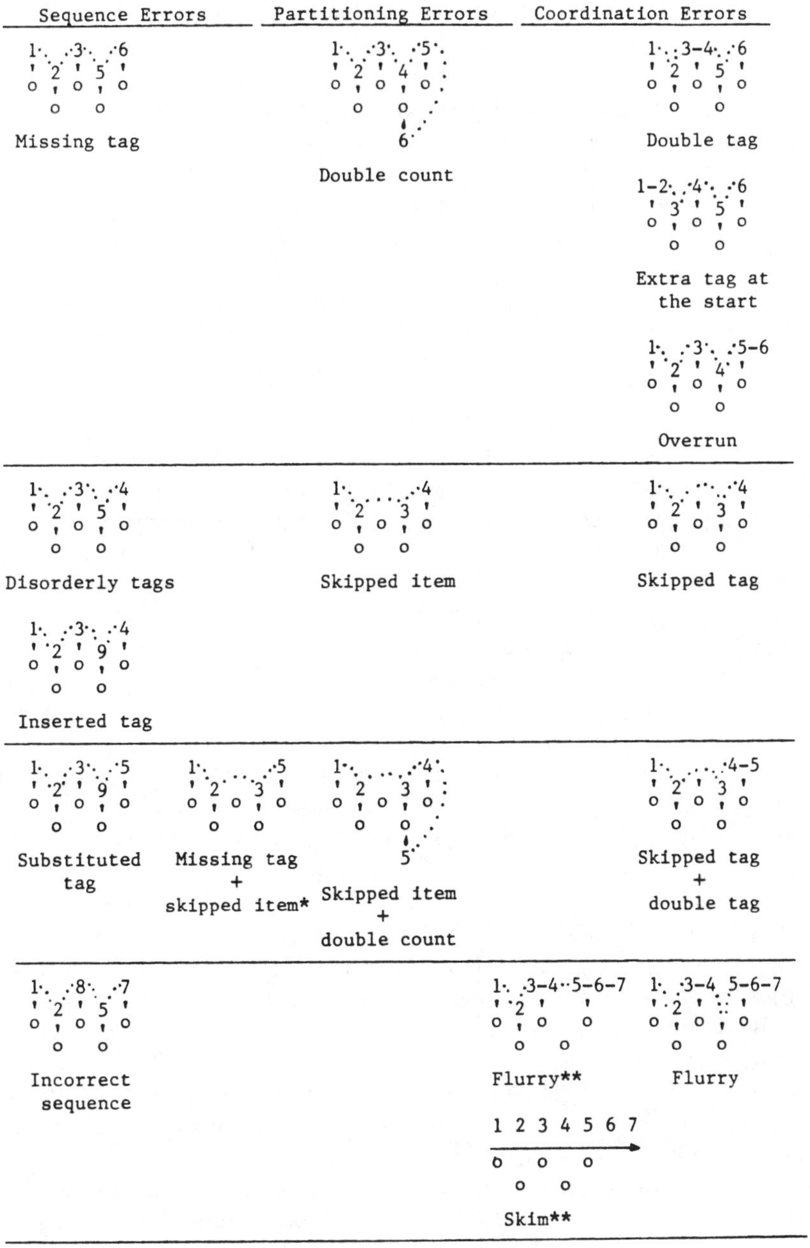

Sequence Errors	Partitioning Errors	Coordination Errors
Missing tag	Double count	Double tag
		Extra tag at the start
		Overrun
Disorderly tags	Skipped item	Skipped tag
Inserted tag		
Substituted tag; Missing tag + skipped item*	Skipped item + double count	Skipped tag + double tag
Incorrect sequence		Flurry**; Flurry; Skim**

' indicates a pointing action.
* denotes a combination of sequence and partitioning errors.
** denotes a combination of partitioning and coordination errors.

nate oral counting and pointing (a coordination error), or both (see Figure 6.1). With skims no effort is made to keep track or coordinate the number sequence with pointing to each item.

Especially with children who simply "skim," enumeration training should emphasize (a) counting slowly and carefully, (b) applying one tag to each item, (c) pointing to each item once and only once, and (d) organizing the count to minimize the effort of keeping track. For fixed items, keeping track of counted and yet-to-be-counted objects is facilitated by learning strategies like starting in a well-defined place and proceeding systematically in one direction (e.g., from left to right). An organizing strategy for counting movable items is to push the counted item aside so that it is clearly separate from the "to count" pile.

CARDINALITY RULE. By the time they enter kindergarten, children routinely apply the cardinality rule for even larger sets (Fuson, Pergament, Lyons, & Hall, 1985). A kindergartner who does not is at great risk. Though many mentally retarded children will spontaneously learn the cardinality rule, many others may need explicit instruction. If a child simply guesses at the cardinal value of a counted set or reenumerates the set, explain the cardinality rule to the child: "When you count, remember the last number you say, because it tells us how many things were counted." If a child reiterates the whole number sequence used in the enumeration process, point out that there is a shortcut: "Let me show you an easy way. After counting, just repeat the last number you said to tell me how many things you counted." It is sometimes helpful if the teacher models the process while "thinking out loud": "How many fingers do I have up? Let me count to see. One, two, three, *four*. Oh, the last number I said was four, so I have *four* fingers up."

CARDINAL-COUNT RULE. Children beginning school typically take even this more advanced cardinality notion for granted; many special education children may not (Baroody & Mason, 1984). This rule can be taught by a two-part procedure developed by Secada, Fuson, and Hall (1983) (see Figure 6.2). The first part involves giving the child a set and indicating (orally and with a written numeral) the cardinal designation of the set. The teacher asks the child to count the set and notes that the outcome of the count was the same as the cardinal designation. For the second part, the teacher puts out another set. The child is again given the cardinal designation and encouraged to count. However, before the child finishes the count, the teacher asks the child to predict the outcome.

PRODUCTION OF SETS. Children typically enter kindergarten with

FIGURE 6.2: Teaching the Cardinal-Count Rule

Part A—Step 1

Teacher: "Here are five circles. [Put out five circles and a card with the numeral 5.] Count the circles to see how many there are."

[5]

○○○○○

Part A—Step 2

[5]

○○○○○

Child: "1, 2, 3, 4, 5"

Teacher: "So, I gave you five circles [point to the numeral card], and when you counted them, the last number you said was five. The number of circles there are is always the same as the last number you say when you count them."

Part B—Step 1

Teacher: "Here are four squares. Count the squares to see how many there are."

[4]

□□□□

Part B—Step 2

[4]

□□□□

Child: "1, 2-"

Teacher: "What's the last number you'll say when you finish counting?" (Teacher corrects and continues as necessary.)

---

an ability to produce at least small sets accurately. A child who is unable to produce up to five objects upon request needs intensive remedial work. Many mentally handicapped children have production difficulties (Baroody & Ginsburg, 1984; Baroody & Snyder, 1983; Spradlin, Cotter, Stevens, & Friedman, 1974) and will need remedial instruction.

A common production difficulty is the "no-stop error": not stopping the counting process after reaching the target. For example, Matt, a mentally handicapped child, was shown eight pencils and told, "Take five for your teacher—remember, just count out five." However, he simply counted the eight pencils. Such no-stop errors have been attributed to memory failure (see, e.g., Resnick & Ford, 1981). According to one memory hypothesis, children simply do not have the target in working memory; they fail to register the requested amount. Or, because they become so preoccupied with the counting process, children forget the target. For instance, asked how many he was supposed to take, Matt indicated, "I don't know." Because he did not record or hold the target in working memory, Matt simply counted all the pencils put before him.

Like many young children (see Flavell, 1970), Matt might have known that a special effort is often required to memorize information—that sometimes we need to rehearse or repeat information to facilitate remembering. For such a child, remedial work needs to emphasize the importance of remembering the target and, if need be, how to remember the goal of the task. Encourage the child to rehearse (repeat) the target to ensure that it is registered firmly in working memory before counting objects. If need be, the child can be encouraged to record the numeral before counting out objects.

Toddlers (Wagner & Walters, 1982) and some mentally handicapped children (Baroody & Ginsburg, 1984) have production problems even when it appears they remember the target. Fred, for example, was asked to count out three objects from a pile of five items but simply counted all the objects: "1, 3, 4, 6, 11, [then, pointing to the last item again] 3." Apparently because he remembered the target, this mentally handicapped lad then retagged the last item "three." Asked to produce five items from a set of nine, he again made a no-stop error but ended the count with the correct tag: "1, 2, 3, 4, 5, 6, 8, 9, 5." Though he failed to stop when the target tag was first produced, Fred appeared to remember the goal and so made sure the last item was given the appropriate tag.

This "end-with-the-target" error can be explained by a second memory hypothesis. Though they register and can later recall the target, the process of counting objects absorbs so much attention some children may fail to match the number sequence of the counting-out process against the target. Because his working memory was so taxed by the counting-out process,

Fred might not have been able to attend simultaneously to a counting and a matching process. Once freed from attending to the counting process, Fred could then recall the target and amend his count.

When a child has no problem recalling the target, remedial efforts should focus on the matching process. Have the child record the target. Then put out (or have the child put out) the first item. Ask the child (pointing to the recorded numeral if that is needed): "Is this the right amount? Should I [you] stop here?" Proceed until the requested amount is achieved. Make explicit why you stop the counting process: "We stop at *N* [say the target number] because *N* [pointing to the target] is how much we need." Especially at first, help the child to find ways of making the counting-out process as easy as possible. For example, simplify the process of keeping track of counted and uncounted items by pushing counted objects away into a clearly separate pile.

Yet another explanation for production errors is that very young and some mentally handicapped schoolchildren do not have a conceptual basis for understanding the task. Children who do not understand the cardinal-count notion may not realize that they are supposed to match their count against the target. To remedy production difficulties then, a teacher should first ensure that a child has the prerequisite cardinal-count skill (Baroody & Mason, 1984).

Magnitude Comparison

By the time they enter kindergarten, nearly all children can make gross and small-number (one to five) number-neighbor comparisons, and a majority will have mastered number-neighbor comparisons up to 10. Primary-level special education and many intermediate-level mentally handicapped children may have deficiencies with even gross and small-number number-neighbor comparisons. Remedial work should begin with concrete objects and with familiar numbers that are obviously different in magnitude (gross comparisons between one, two, or three and larger numbers like nine or ten; number-neighbor comparisons between one and two or two and three).

A variety of games can be easily assembled that will involve concrete models (see Example 6.3). For example, in the *Moon Invaders* game, players compare the length or height of two sets of interlocking cubes. Thus the comparison of numbers is connected to and reinforced by obvious perceptual clues: "You have eight spaceships on the moon; I have two. See how *big* your row of spaceships is. Eight spaceships is *more* than two." Gradually the child will learn the ideas that numbers are associated with magnitude and that numbers coming later in the number sequence are more. Once

EXAMPLE 6.3

Concrete Number-Comparison Games

MOON INVADERS

Objective: Gross or number-neighbor comparisons of numbers 1 to 10

Materials: (1) A number of differently colored moons (paper circles)
(2) Two differently colored sets of interlocking cubes and
(3) Spinner with numbers 1 to 10 (for gross comparisons) or a set of index cards listing specific comparisons for either objective)

Instructions: Lay the circles out on the table. Give one set of cubes to each of the two players. Explain that the circles are moons and the cubes are spaceships. Whoever lands more spaceships on a moon wins it, and the player who captures the most moons wins the game. Use the spinner or cards to determine the number of spaceships each player may land. Ask a child which player has landed more, for example: "You have five spaceships and Billy has three spaceships. Which is more, five or three?" If needed, point out the different lengths (or heights) of the two sets of interlocking cubes.

DOMINOES MORE (LESS) THAN

Objective: Number-neighbor (or one-less-than) comparisons of number 1 to 10

Materials: Set of dominoes

Instructions: Based on a game suggested by the Wynroth (1969/1980) curriculum, dominoes more (less) than as played like regular dominoes with one exception. Instead of matching numerically equivalent sets to add on new dominoes, a new domino must have a set one more (less) than the item on the end. The figure below illustrates "Dominoes Less Than." A player is about to add "8" to the end with "9."

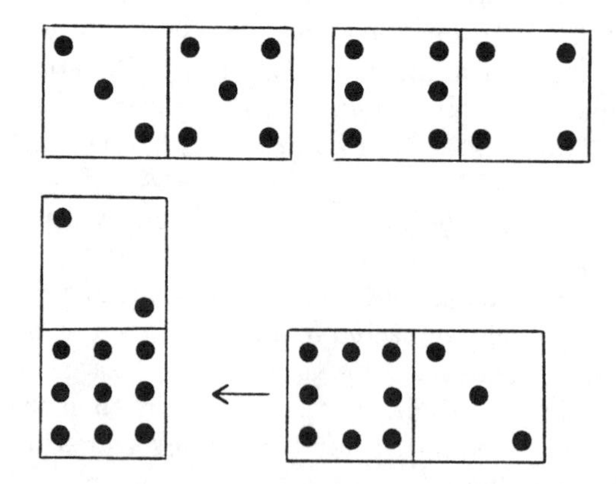

these basic ideas take hold, the child should be weaned from activities that rely on concrete objects and required to solve the problems mentally.

With special education children in particular, it may be helpful to point out the counting strategy that can be used to compare number neighbors and how this strategy is linked to the prerequisite number-after skills. Explain, "To figure out which number is bigger, let's count and see which number comes after. For 4 or 3, we count: '1, 2, 3' and after 3 is 4, so 4 is more." It may help to model the procedure for the child and to use a number list or interlocking blocks to count on. Eventually, the counting procedure can be faded and the child can be asked, "Which is more, 4 or 3; which number comes *after* when we count?" Another way to make the connection between the comparison and number-after skill explicit is to follow up number-after questions with which-is-more questions. For instance, "What comes right after 3 when we count? We say 3 and then . . . ?" After the child responds, ask: "So which is more, 3 or 4?" (Note that to force the child to really think about the comparison, the larger number is mentioned first or "out of the usual order" about half the time.)

EDUCATIONAL IMPLICATIONS: TEACHING COUNTING SKILLS

Some general instructional guidelines are summarized below.

1. *Children should master each counting skill to the point where it is automatic.* This is essential because count skills build on one another and serve as the basis for other complex skills, such as counting out sums or making change. If basic skills are not efficient, they cannot be effectively integrated with other skills to perform more complex functions.

2. *Remedial efforts should build on concrete experiences.* To be meaningful, basic counting-skill instruction should be grounded in concrete activities. Moreover, especially with special education populations, it may be important to explicitly link concrete activities to the skill being taught.

3. *Remedial efforts should, over an extended period of time, provide regular practice with activities of interest to the child.* Incomplete mastery of basic counting skills can usually be attributed to a lack of experience or interest. Unless practice is made interesting, some children will not engage in the extended and involved experience necessary for skill mastery. For example, children very quickly grow tired of oral-counting drills. Because it is meaningful, children are more willing to generate the number sequence in the context of enumerating objects (Fuson et al., 1982). The form practice should take depends on the child. Many children will re-

spond enthusiastically to various types of games that involve counting; some children might prefer play with a muppet/puppet; and other children may just enjoy the one-to-one contact with an interested and enthusiastic adult or child tutor. The key point is that the practice need not—indeed should not—be uninteresting to the child.

Additional games and activities for teaching oral-counting, numbering, and magnitude-comparison skills are detailed below.

Games and Activities

HIDDEN STARS

Objectives: (1) Enumeration
(2) Cardinality rule

Materials: Cards with stars or other pictured objects (1 to 5 for beginners)

Instructions: Explain: "We're going to play the hidden-stars game. I will show you a card with stars on it. You count the stars. After you finish counting, I will hide the stars. If you can tell me how many stars I am hiding, you win the point." Flip over the first card and have the child count the stars. Cover them with your hand or a piece of cardboard and ask, "How many stars am I hiding?" The child should respond with just the cardinal value of the set. If the child counts from one, ask if there is an easier way of indicating how many stars were counted. If necessary, teach the child the cardinality rule directly by modeling the task and "thinking out loud" (describing the procedure and rationale).

COUNT PREDICTION

Objective: Cardinal-count concept

Materials: Small countable objects such as blocks or poker chips

Instructions: Give the child a set of blocks (for example, 5), and say, "Here are 5 blocks. How many will we get if you count them?" Afterward, have the child count the set to check. This can also be done with a die. After a roll, don't allow the child to count count the dots immediately. Instead follow the procedure described above.

CAR RACE

Objectives: (1) Enumeration
(2) Production

Materials: (1) Race-track gameboard (a row of squares that spirals inward)
(2) Die (with 0 to 5 dots initially; 5 to 10 for more advanced children)
(3) Small matchbox-type cars

Instructions: Have the children choose the cars they would like to use. Put the cars at the beginning of the path. Take turns rolling the die and moving the cars that number of spaces on the path. Have the players count both the dots on the die (enumeration) and the spaces when they move (production). These skills can also be practiced by playing other basic board games with various themes according to the child's interests.

FILL-IN

Objectives: (1) Enumeration
(2) Production

Materials: (1) Individual pegboards or race tracks
(2) Pegs or chips
(3) Deck of cards with dots (1 to 5 for beginners; 6 to 10 for more advanced children)
(4) Small trays (e.g., plastic lids)

Instructions: Give each child a pegboard or spiral race track. Say, "We're going to see who can fill up their pegboard (race track) first." Have the children take turns picking a card from the deck and counting the dots to determine how many pegs (chips) to take. Tell the child to take this amount. Have the child count out the pegs (chips) onto the small tray. (This procedure makes correcting production errors less confusing.) If a mistake is made, empty the tray. Have the child try again or, if necessary, help the child produce the correct number. After the correct number is produced, have the child place the pegs (chips) on his or her gameboard. The child who fills in his or her spiral first is the winner.

TURN OVER

Objective: Number just after or before a given number (1 to 9)

Materials: Numeral cards 1 to 9

Instructions: The basic version of this game is described in more detail in Bley and Thornton (1981), along with other games such as Walk On and Peek that are useful in teaching number after. For the basic version of Turn Over, lay out the numeral cards, face up, in order on table. Tell the child to hide his eyes, then turn a card face down, and have the child uncover his eyes. Then he may look and figure out which card is turned

CARL A. RUDISILL LIBRARY
LENOIR-RHYNE COLLEGE

over. Point to the card before (after) the covered card and say, for example, "What's this card? What comes just after [before] 6?" Continue until each number has been turned over once. The basic version is especially useful with children who cannot use a running start (count from one) to answer number-after questions and for children who confuse number before with number after. A more advanced version involves eliminating the visible clues of the number sequence and requires the child to solve the problems mentally. Simply lay the cards out face down. Then turn over one card and ask the child what comes next (before).

NUMBER-SQUARE RACE

Objective: Gross comparisons of numbers 1 to 10
Materials: (1) Number-square track (about 6" x 30") with the numbers 1 to 10 ruled off (see Figure 6.3)
(2) Matchbox-type cars
Instructions: Let the players choose the cars they would like. Place the cars at the starting line—about 6 inches left of "one" on the number-square track. Tell the children that their cars are going to race and the car that goes *farthest* will win. Have the children give their cars a push down the track. Cars that go off the side or the end of the track are disqualified. If a car lands on the line between numbers, the car is placed on the space where the greatest portion of the car rests. After both players have pushed their cars, ask one: "Your car went 5 and Jane's car went 3. Which is more—5 or 3? Who wins?" Vary the order in which the numbers are presented so the larger number is sometimes first and sometimes last. If necessary, correct the child by showing on the number list that the bigger number involves moving more spaces.

CHASE GAME

Objective: Number-neighbor comparisons
Materials: (1) Spiral race-track gameboard
(2) Two tokens (plastic figures)

FIGURE 6.3: Number-Square Track

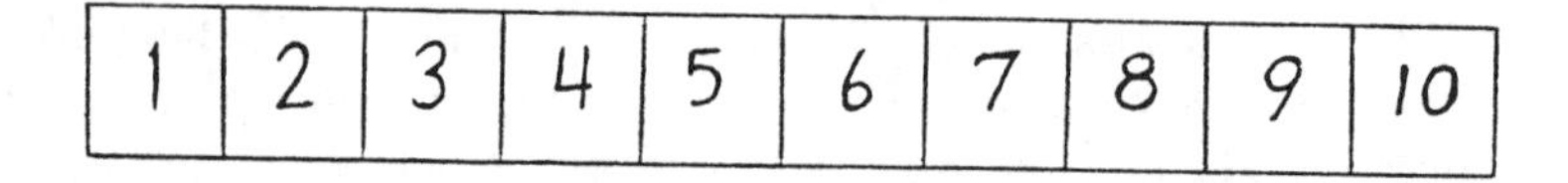

(3) Index cards with different comparisons (numbers 1 to 5 for beginners; larger numbers for more advanced children)

Instructions: Tell the child that your player (token) is going to chase his or hers around the board. Draw a card and read two numbers listed. Encourage the child to pick the bigger number. The child's choice indicates how many spaces his or her token advances; the unchosen number, how many spaces you can move your token. After each turn, note the tokens' positions, for example: "Yes, that's the one with more. Your token is still ahead," or "No, that's not more. Look, my token is catching up with yours." If the child has difficulty, a number line or blocks can be used for illustration.

SUMMARY

Orally generating the number sequence is only a first step in mastering a complex of important skills, which adults use routinely and automatically. By the time they enter school, children typically can generate the rote portion of the number sequence and some of the rule-governed portion, enumerate and produce sets of objects, use the cardinality rule to summarize their enumeration, and even use number-sequence (number-after and number-before) relationships to determine the larger of two quantities. Some children, especially mentally handicapped children, may need remedial work to master these basic informal skills. In the first years of school, children solve the decade problem and extend their oral counting to 100 and beyond. As the number sequence becomes more familiar, they learn skip counts, such as counting by twos, and how to count backwards. Initial or remedial instruction should ensure that each successive component in the hierarchy of counting skills is mastered. Training should be concrete, extended, and interesting.

CHAPTER 7

Number Development

Does an ability to understand and use number develop directly from children's counting experience? Or must the development of meaningful counting await the acquisition of prerequisite concepts or capacities? What can a child learn about number from counting experiences? What role does pattern recognition play in mathematical development? Is the cardinal (set-theory) approach of the New Math or the logical training of Piagetian programs useful with young children? What roles should counting experiences play in teaching children numerical concepts?

TWO VIEWS OF NUMBER DEVELOPMENT

Conservation Problems: The Case of Peter

Peter, a preschooler, spread out seven blue chips in a row in front of him. I put out a row of seven white chips in one-to-one correspondence and, while Peter watched, added an eighth white chip. The row of eight white chips was then shortened. Peter was then encouraged to count to see if there were the same number of chips in each row or if one row had more. Peter responded, "My row has [counts the blue chips] 1, 2, 3, 4, 5, 6, 7. Your row has [counts the white chips] 1, 2, 3, 4, 5, 6, 7, 8. See, your row *only* has eight—my row has more!"

Despite counting both sets, Peter still responded to the conservation-of-inequivalence question incorrectly. Apparently, oral counting and enumeration ability do not necessarily imply a well-developed understanding of number. Why did counting not help Peter, and what kind of instruction would improve Peter's understanding of number?

The Logical-Prerequisites View

Psychologists offer two different explanations for how number words and the act of counting take on meaning. According to one view, before reaching the "age of reason" (seven years or so), children are incapable of understanding number and arithmetic (e.g., Piaget, 1965). Peter's curious response is attributed to an inability to think logically. That is, Peter presumably lacks the logical concepts and reasoning necessary for a number concept and meaningful counting. Because counting does not ensure success on equivalence- or inequivalence-conservation tasks, some psychologists (e.g., Wohlwill & Lowe, 1962) have concluded that counting experience has little, if anything, to do with the development of a number concept. For example, Piaget (1965) argued that children learn to recite the number sequence and arithmetic facts at a very young age and that these are entirely verbal and meaningless acts. Even enumeration does not guarantee an understanding of number. In this view, the developments of a number concept and meaningful counting depend upon the evolution of logical thought.

THE CARDINAL MODEL. According to one logical-prerequisite model, children must understand classification before they can comprehend the essential meaning of number. This involves learning how to define a set—to classify objects so that each can be assigned to a correct set. For instance, a set of curved shapes could include c, C, u, U, s, S, and O but not l, L, v, V, I, F, and #.

Understanding class logic also requires comprehending hierarchial classification or "class inclusion": A class is the sum of its parts (subclasses) and, therefore, is greater than any subclass. For example, given three roses and five violets and asked, "Are there more violets or more flowers?" a child should respond that the class (flowers) has more than the subclass (violets). Yet young children have trouble with such class-inclusion questions (e.g., Piaget, 1965). Such results have been taken as evidence that young children do not appreciate class logic and hence are incapable of truly understanding number.

Furthermore, class logic entails understanding the idea of equivalent sets. The equivalence of sets is defined by one-to-one correspondence: Two sets belong to the same class if the elements of each can be matched up. Equivalence and one-to-one correspondence, which are a foundation of formal mathematics, are considered the psychological foundation of mathematical learning.

THE PIAGETIAN MODEL. According to Piaget (e.g., 1965), children must understand the logic of relations (serial ordering) as well as classes to comprehend equivalence relationships and hence the meaning of number. Piaget agreed that equivalence (one-to-one correspondence) is the psychological foundation for understanding number. However, Piaget felt that understanding one-to-one correspondence involved understanding *both* classes and serial ordering. For example, matching involves noting the first element of each set, the second, the third, and so forth. In other words, to establish a match, children have to keep track of which item they have matched up by imposing an order.

Likewise, he viewed number as the union of serial-ordering and class concepts. For example, enumerating a set involves treating all the items of the set as members of the same class yet within the set distinguishing between the first item, the second item, and so forth. Moreover, numbers follow in order and form a hierarchy of classes. For example, three is a class that contains the subclasses one and two (and is itself a subclass of larger numbers). In summary, Piaget argued that number could not be understood in terms of a single logical concept but is a unique synthesis of logical ideas (Sinclair & Sinclair, in press).

For Piaget (1965), the development of an understanding of number and meaningful counting is tied to the onset of a more advanced stage in thinking ability. The logical prerequisites of number (concepts of serial ordering, classification, and one-to-one correspondence) appear with the "operational stage" of mental development. Children who have not achieved the operational stage cannot understand number or engage in meaningful counting, whereas operational children can. Number then is an all-or-nothing concept.

Piaget (1965) argued that number conservation was of the utmost importance because it signaled the achievement of the operational stage: the advent of logical thinking; an understanding of classes, relations, and one-to-one correspondence; a true number concept; and meaningful counting. More specifically, Piaget interpreted number conservation as indicating an understanding that, once the equivalence (inequivalence) of two sets has been established, changes in the configuration of the sets do not change the equivalence (inequivalence) relationship. That is, equivalence (inequivalence) relationships are *conserved* over irrelevant transformations in the physical appearance of a set. A "conserver" realizes that the number of items in a set does not change because the appearance of the set has changed.

The Counting View

An alternative view is that Peter's difficulty with the conservation task is the result of incomplete counting knowledge, not of a complete inability to think logically. Some psychologists (e.g., Gelman, 1972; Zimiles, 1963) have concluded that counting is key to the development of children's understanding of number. Number is not seen as an all-or-nothing concept made possible by a general change in the way children think (a new stage of mental development). Instead, the counting model proposes that an understanding of number evolves slowly as a direct result of counting experiences.

In this view, a number concept and meaningful counting develop gradually, step by step, the result of applying more and more sophisticated counting skills and concepts. Preschoolers often first learn to use numbers mechanically and then gradually discover or construct deeper and deeper meanings of number and counting (e.g., Baroody & Ginsburg, in press; Fuson & Hall, 1983; von Glasersfeld, 1982; Wagner & Walters, 1982). As their understanding of number and counting increases, children apply number and counting procedures in increasingly sophisticated ways. More sophisticated procedures and applications, in turn, lead to further insights, and so forth. In effect, the development of skills and concepts is intertwined. Indeed, in recent years, some Piagetians (e.g., Elkind, 1964; Piaget, 1977; Sinclair & Sinclair, in press) have concluded that an analysis of number development would be psychologically incomplete without taking into account the contribution of counting activity.

Counting Concepts

Initially, children merely chant number words. At this point counting appears to be nothing more than a meaningless "sing-song" (Ginsburg, 1982). For example, 22-month-old Arianne chants "two, five, two, five" as she hops down four steps. She has heard her brother and sister, 3-year-old twins, reciting numbers while hopping down stairs and engaging in other play. It appears that Arianne has learned that certain activities can be accompanied by reciting number words. She is imitating the procedure (and only a part of the correct number sequence) modeled by her siblings. "Number words are words and, as happens with other words, children can learn to say them long before they have formed [mental images], let alone abstract concepts to associate with them . . . " (von Glasersfeld, 1982, p. 196).

At first, children may engage in enumeration without an intent to number sets. For example, 2-year-old Arianne just seems to enjoy the act of tagging objects as she sorts through her playthings; no effort is made to use one tag for each item or to summarize the count. When asked how-many questions, she realizes that the correct procedure entails responding with a number, but she does not yet seem to appreciate that numbers are used to designate the cardinal value of a set and to differentiate a set from other sets of a different cardinal value. Consider the following exchange between Arianne and her father:

FATHER: [Turns to picture with two cats.] How many cats are in the picture?

ARIANNE: Two.
FATHER: [Turns to picture with three dogs.] How many dogs are in this picture?
ARIANNE: Two.
FATHER: [Turns to a picture with one cat.] How many?
ARIANNE: Two.

"Two," it appears, is Arianne's stock answer for responding to a how-many question. At this point, counting is entirely a verbal act without meaning.

Note, though, that she already treats numbers as a special class of words. Only numbers are used when she is asked how many or when she is asked to count. Children appear to distinguish between counting and noncounting words very early (Fuson et al., 1982). For example, preschoolers only very rarely use letters when asked to count (e.g., Gelman & Gallistel, 1978). Even moderately mentally handicapped children at the elementary level nearly always recognize numbers as a special class of words that are applicable to counting activities (Baroody & Ginsburg, 1984).

STABLE-ORDER PRINCIPLE. In time, as children use and reflect on their counting skills, they learn or discover important regularities about their counting actions and numbers. Children appear to learn the first number-sequence terms by rote. Initially, they may not use the same terms and order of terms when they chant numbers or count objects. For example, 3-year-old Alexi did not always start with one to count sets. At some point, children realize implicitly, or even explicitly, that counting requires producing the same order of number words every time. A stable-order principle specifies that a consistent sequence of number words is a necessary condition for counting. Children whose actions are guided by a stable-order principle may use the conventional number sequence or their own idiosyncratic (nonconventional) sequence—but they do so consistently (Gelman & Gallistel, 1978). For example, Beth always uses the correct chain "1 to 10," and Carol always uses her own version ("1, 2, 3, 4, 5, *6*, *8*, 9, 10, *18*") to count 10 objects.

ONE-TO-ONE PRINCIPLE. As a result of imitation, children may initially—like Arianne—chant numbers as they point to objects and may even develop some enumeration proficiency with small sets. Later, they may realize the necessity of tagging each item in a set once and only once. A one-to-one principle underlies a genuine effort to enumerate sets and guides efforts to construct keeping-track strategies, such as pushing counted items into a pile away from uncounted items. Children as young as three appear to use such a principle to detect enumeration errors, such as double counts or skips (Gelman & Meck, in press).

UNIQUENESS PRINCIPLE. Because one function of counting is to assign cardinal values to sets so as to distinguish between them or compare them, it is important that children not only generate a stable sequence and assign one and only one tag to each element of a set but also use a sequence of distinct or unique tags. For example, a child might use the sequence "1, 2, 3, 3" consistently and use these tags in a one-to-one fashion but, because each item is not distinct, the child will label sets of three and four objects the same (with the cardinal designation "3") (Baroody & Price, 1983). Even when a child has to resort to using nonconventional terms, an appreciation of a uniqueness principle (an understanding of the differentiating function of counting) would preclude choosing previously used terms. For instance, consistent use of the nonconventional sequence "1, 2, 3, eleventeen" would mislabel sets of four objects but at least differentiate sets of four from smaller sets. Thus, in addition to stable-order and one-to-one principles, it is important for children to adhere to a uniqueness principle.

ABSTRACTION PRINCIPLE. Children also have to learn how to define a set for counting purposes. The abstraction principle addresses the issues of what can be grouped together to form a set (Gelman & Gallistel, 1978). For counting purposes, a set can be made up of like objects (e.g., balls: • • •) or unlike objects (e.g., balls, stars, and sticks: • * –). To include unlike items in a set, all a child has to do is overlook the physical differences of the items and classify each item as a "thing" (e.g., a ball, star, and block can be thought of as one, two, three *things*). In effect, to create a set of unlike items, we find (abstract) something common to all the items.

CARDINALITY PRINCIPLE. By imitation, children can readily learn the counting skill called the cardinality rule: stating the last number counted in response to how-many questions. However, use of a cardinality rule does not guarantee a deep appreciation of cardinality (Fuson & Hall, 1983; von Glasersfeld, 1982). It does not necessarily mean that the child realizes that the last term designates the set's quantity and that a set will have the same label if it is recounted after, say, spreading out the objects. For example, one mentally handicapped boy correctly used a one-to-one procedure to enumerate 15 objects but used the number sequences "1, . . . 5, 19, 14, 12, 10, 9, 20, 49, 1, 2, 3" (Baroody & Ginsburg, 1984). In response to the how-many question, he was content to reply, "three!" Apparently the notion of "threeness" did not exclude sets five times greater than three!

Children may construct a cardinality principle by reflecting on their counting activities. By counting a collection of, say, three toys, spreading the toys out, and recounting the toys, a child can discover that a collection retains the same (cardinal) designation ("three") regardless of appearance.

ORDER-IRRELEVANCE PRINCIPLE. It appears that an order-irrelevance principle ("The order in which a set is enumerated does not affect the cardinal designation of the set") is also discovered by reflecting on counting activity (Baroody, 1984d). Consider a case described by Piaget (1964). The boy—about 4 or 5 years old at the time—counted 10 pebbles in a row. Because he did not realize the outcome would be the same, the child counted the pebbles in the opposite direction. Again he found 10. Interested by this result, the child arranged the pebbles in a circle, counted them, and once more arrived at 10. Finally, the boy counted the circle of pebbles in the opposite direction only to find the same result. By counting the elements in various ways, the boy discovered an interesting property about counting actions: The arrangement of elements and the order in which they are enumerated were irrelevant in determining the cardinal designation of the set.

Concepts of Equivalence, Inequivalence, and Magnitude

Once children have mastered these counting concepts, which pertain to a single set, counting can be used in the more complicated contexts of comparing two sets. Counting can also be used to discover that appearance is irrelevant in determining whether or not two sets are equal. If a child counts two sets and the resulting numbers are identical, the child can conclude that the sets have the same number of objects—despite differences in appearance. Children probably discover this fundamental number notion by playing with small sets of one to four objects. For example, children may label as "two" various pairs of things (e.g., blocks or fingers) including natural pairs of things (e.g., eyes, arms, twins). Because the child can readily *see* that such differently composed sets match up, they eventually conclude that sets labeled "two" are equivalent in number despite differences in physical appearance (e.g., Schaeffer et al., 1974). This understanding is then applied to larger sets that a child cannot readily match visually or mentally.

Before school, children also learn that number can specify differences in sets (inequivalence) and be used to specify "more" or "less" (order sets according to magnitude). Again, this probably comes from playing with small sets of objects. For example, a child might have a choice of three baskets with one, two, or three candies. The child can readily see that three is more than one or two, and that two is more than one. By counting each set, number labels are associated with these perceptible differences in magnitude. A child might also count, say, two blocks ("one, two—two blocks"), then add one more, and conclude that there is "more." The child may again count the blocks ("one, two, three—three blocks!") and find that the number label is now "three." From repeated instances of both kinds of concrete experience, a child might conclude that (a) different numbers are associated with different magnitudes, (b) the larger of two numbers always comes later in the count sequence, and (c) each count term is one more than the preceding term in the number sequence.

Finger counting may play a key role in such number development. By counting on their fingers (extending fingers as they say "one, two, three . . ."), children can *see* the number of fingers getting larger as they count higher. In this way, children may recognize that magnitude is associated with position in the number sequence. By counting on their fingers, they may even realize that two is one more (finger) than one, three is one more (finger) than two, and so forth. In brief, as a result of counting experience with small sets and their fingers, children may learn number-based rules for judging "same number," "different amounts," and "more."

CONSERVATION OF NUMBER. In time, number-based rules for evaluating equivalence, inequivalence, and magnitude permit children to conserve. These accurate number-based criteria free children from depending on perceptual cues, such as length, when making quantitative comparisons. As a result, children are no longer misled when a row is lengthened or shortened during a number-conservation task. Paul, who concluded that his longer row of seven had more than a shorter row of eight, may not have sufficient counting experience to accurately compare number neighbors. That is, this preschooler may not have learned number-based methods or skills for gauging the relative magnitude of two relatively large sets.

Even after children have learned number-based rules for determining equivalence or inequivalence and making magnitude comparisons, they may fail to use these rules on a number-conservation task for a number of reasons. First, they may not think to count and thus have no basis for using number-based rules. After one row is physically transformed (e.g., lengthened), children may be unsure about the initial relationship of the sets (maybe the rows were not equal to begin with). Given this uncertainty, they may be overpowered by the visual cues of the uneven rows, may seize upon a perceptual criterion of length, and conclude that the longer row has more (Acredolo, 1982). Nonconservers, then, may not actually believe that lengthening a row adds something to the row. Moreover, nonconservation is a logical contradiction only if one believed that the rows were equivalent initially, which without counting and specific numbers is an uncertain proposition for young children. Nonconservation does not necessarily imply that a child could not reason logically about equivalence relationships if the child counted and used numbers (Gelman & Gallistel, 1978).

Second, even if they think to count, young children may not have sufficient confidence in their number-based rules to rely on a numerical rather than a perceptual criterion (e.g., Gelman, 1972). The number-conservation task sets up a conflict between a child's existing rule for comparing numerosities ("If one row is longer than another, it has 'more'") and the developing counting-related rule ("If two rows are counted and have the same number label, they are equally numerous"). A young child may resolve the conflict by simply falling back on the familiar perceptual criterion. A somewhat more experienced counter may be torn between the two criteria and respond inconsistently.

Eventually children resolve the conflict by devising a new and more sophisticated rule that integrates the perception-based and number-based rules. In effect, the new rule specifies, "If one row is longer than another, it may be more numerous—*unless* counting produces the same number label, in which case the rows are equally numerous." Basically, children seem to resolve the cognitive conflict by reorganizing the existing information into a more systematic form. In this way, children may continue to use perceptual cues where differences are clear (e.g., choosing six candies as opposed to two) (Zimiles, 1963). In cases where the difference is unclear (e.g., two movie lines, where one is long but spread out and the other is short but densely packed), the rule suggests a need to count and make a number-based judgment.

Older children do not even have to count to conserve. They take number conservation for granted. Indeed, they think it is odd that an adult should ask about the obvious. From repeated counting experiences, they know that if nothing is added or subtracted, matching sets remain equivalent regardless of appearance (Lawson, Baron, & Siegel, 1974). That is, children eventually induce a relatively *abstract* equivalence rule based on one-to-one correspondence (matching), which supplements their more concrete equivalence rules based on specific numbers (Gelman & Gallistel, 1978).

In fact, there is considerable evidence that abstract equivalence/inequivalence rules evolve out of children's concrete counting experience. Young children frequently use counting as the basis for their number-conservation judgment (e.g., Gelman, 1972). Furthermore, the teaching or development of accurate numbering skills facilitates the acquisition of number conservation (Bearison, 1969; LaPointe & O'Donnell, 1974; Starkey & Cooper, 1977). Indeed, young children often appear to go through a stage where they rely on counting to conserve (conservation with "empirical verification") before they conserve by insight (conservation with "logical certainty") (Apostel, Mays, Morf, & Piaget, 1957; Greco, Grize, Papert, & Piaget, 1960; Green & Laxon, 1970).

According to the counting view, then, counting experience is the key for making explicit and extending intuitive notions of equivalence, inequivalence, and magnitude order (Baroody & White, 1983). As we saw in Chapter 2, even infants can visually inspect and intuitively determine whether small sets (sets of up to four objects) are equivalent or not. Counting provides verbal labels that can be appended to these small sets. It is counting experience that provides the basis for formulating explicit, number-based rules and then more abstract (matching-based) rules for reasoning about numerical relationships between larger quantities. Thus, children typically will first depend on counting to figure out equivalence relationships, such as that represented by number conservation, and only later depend on relatively abstract rules based on matching. In brief, counting rather than matching seems to be children's natural avenue to understanding equivalence, inequivalence, and magnitude relationships with nonintuitive numbers.

Basic Arithmetic Concepts

Through counting experiences, children also discover what does change number. If changes in order or arrangement do not affect the cardinal value of a set, certain types of transformation do change number (e.g., adding and subtracting objects). Once children become proficient in enumeration or can subitize number patterns, they are ready to notice important arithmetic relationships. A child can readily determine or see that one block added to another is "two" and that one added to two makes "three," and so on (Baroody & White, 1983; Ginsburg & Baroody, 1983; von Glasersfeld, 1982). Similarly, a child can readily determine or see that one cookie taken away from three leaves two. There is only a thin line between counting and incrementing or decrementing by one. Discovering the effects of adding or subtracting one depends upon efficient numbering skills.

From their informal counting experiences, children construct basic but general arithmetic concepts. Specifically, as a result of informal experiences, children view addition as an incrementing process (as adding something more to a given quantity) and subtraction as a decrementing process (as taking something from a given quantity). For example, at the beginning of kindergarten, Aaron was asked how much he thought four and five (4 + 5) was. He replied,"If I had to guess, I'd say four or five. Wait, those *are* the numbers. Six or seven." Because he viewed addition as an incrementing process, Aaron knew that it would not do to give an addend as a sum. Because of his informal concept of addition, Aaron readjusted his mental estimate upward so that it would be at least somewhat larger than five.

Consider also preschoolers' reactions to the "magic-show" task developed by Gelman (Gelman, 1972; Gelman & Gallistel, 1978). The first phase of the task establishes the importance of a particular number. A child is shown two plates with different numbers of plastic toys (e.g., a plate with three mice and a plate with four mice). The tester then points to one plate (e.g., the plate with three mice) and designates it the "winner." Though not instructed to do so, children frequently count or note the number of mice on the plates. The plates are moved behind a screen, covered, shuffled, and re-presented to the child. The child then tries to pick the winner. If the nonwinner is uncovered (e.g., the plate with four mice), the child is given a second pick and, of course, finds the winner. This process is repeated until the child expects to find the winner, if not on the first then certainly on the second pick.

The second phase of the task gauges the child's reaction to various types of transformations. Behind the screen, the experimenter would sometimes make number-irrelevant transformations: change the position of the toys (e.g., change a row of three mice to a triangular form), alter the color of an object, or even substitute a different toy for one of the objects. Sometimes, the tester secretly made number-relevant transformations: added or subtracted toys to the winner (e.g., add one toy mouse to the plate of three so that neither plate was a winner).

The children's reactions to these number-irrelevant and number-relevant transformations were then noted. Children ignored the number-irrelevant transformation—the winner (e.g., "three") was still the winner. However, children were quite surprised when they uncovered both plates and could not find the winner. When asked what happened, children pointed out that something had been added to (taken away from) the winning plate. When asked how to correct the situation, children indicated the extra toy had to be removed (the missing toy had to be put back).

Such response patterns might not seem like much of an achievement to adults, but they suggest some important competencies on the part of preschool children. Despite the fact that a child might not conserve number, success on the magic task implies an understanding of which transformations are and are not important to changing number (e.g., addition and subtraction change numbers, rearrangement does not) at least with familiar numbers. Moreover, they appear to understand that addition and subtraction are inverse operations: One undoes the other. Thus even young nonconservers have some understanding of arithmetic and can, within limits, reason logically about numerical relationships.

The Role of Pattern Recognition

"Subitizing" entails automatically recognizing number patterns (e.g., without counting, identifying •˙• or • ˙ • as "three"). The place of automatic recognition of number patterns in the development of number remains an open issue. Some theorists (e.g., Klahr & Wallace, 1973; von Glasersfeld, 1982) suggest that children can subitize small numbers before they can count. In the Piagetian view, very young children simply recognize a *whole pattern*. For example, •˙• is seen as a global configuration that is associated with "three"; • • • is seen as a different global configuration that just happens also to be associated with "three." Neither "whole" is recognized as a collection of countable items—a collection composed of units (individual elements). In this view, subitizing does not imply number understanding. Children do not recognize number patterns as both a whole (as a unit itself) and a composite of parts (individual units) until they achieve the stage of operational thinking. With this intellectual achievement, a child is capable of viewing number and number patterns as a unit composed of units (e.g., Steffe, von Glasersfeld, Richards, & Cobb, 1983).

Another view is that counting precedes subitizing (Beckmann, 1924). That is, children learn to enumerate collections correctly before they can accurately and quickly recognize sets. Indeed, some evidence (e.g., Baroody & Ginsburg, 1984; Gelman, 1977) suggests that automatic recognition of number patterns often develops after extensive object-counting experience. This may be particularly true for mentally handicapped children (Baroody & Ginsburg, 1984). In this view, even preschoolers can recognize that number and number patterns are both a whole collection and a composite of individual parts—a unit composed of units.

In any case, both models suggest that subitizing is a fundamental skill in the development of children's understanding of number. Once children can automatically recognize a pattern, they can discover important things about number. For example, a child who takes three objects in a triangular array and puts them in a row and who recognizes both •˙• and • • • as instances of "three" may implicitly or explicitly formulate the principle: "The way the marbles are arranged does not change how many marbles I have." Subitizing may also play a central role in learning number-based rules for gauging equivalence. Presented with triangular and linear arrays of three, a child who can immediately recognize both as "three" may infer that sets can have the same number even if they do not look alike (von Glasersfeld, 1982).

EDUCATIONAL IMPLICATIONS: NUMBER DIFFICULTIES AND REMEDIES

Counting Principles

By the time they begin school, children are quite knowledgeable about counting (Gelman & Gallistel, 1978; Gelman & Meck, 1983). Practically all appear to take for granted the various principles that underlie or govern counting: the stable-order, one-to-one, uniqueness, and abstraction principles. Most appear to appreciate even the relatively sophisticated order-irrelevance principle. Such knowledge cannot be taken for granted in very young or mentally handicapped children. For instance, these children may not say numbers in any consistent order. A far more common error is to say the first numbers in the correct order but then "spew" terms without order. For example, a child might consistently begin with "1, 2, 3" but then say "6, 8, 12, 9" one time and follow up with "12, 3, 6, 6" another. Note that the second spew involves repeated terms. "Three" was used previously in the first correct portion, and "six" is used consecutively to end the count. This count clearly violates not only the stable-order principle but the uniqueness principle. (Though spews and repeated terms are inconsistent with a stable-order and uniqueness principle, these errors do not always and *necessarily* indicate that these principles are not known. For instance, children may know the principles but not remember that they have used a term previously.)

If children have not had the opportunity to discover these principles, they should be given abundant counting experiences—especially in the contexts of games or activities of interest. Indeed, it may be helpful to point out these principles directly (e.g., "When we count things, we must make sure to say the numbers in the same way every time," or "When we count things, we must make sure to use a *new* number for each thing we point to"). It might be helpful to discuss stories, such as those in Example 7.1 or those that appear regularly in children's television programs such as "Sesame Street."

Equivalence, Inequivalence, and More Than

Children learn to rely on counting or subitizing to determine "same number" (*equivalence*) and "different number" (*inequivalence*) quite early—at least for small numbers. If children do not spontaneously use num-

EXAMPLE 7.1

Counting Stories

Once and Only Once

Count Disorderly was all excited. He jumped up and down and ran all over his castle. It was almost his birthday, and he was going to have a big party! The cook came to ask him how many people he was inviting so she could make enough food and cake. Count Disorderly took out his guest list and started to count the names on the list. Though he lost track of which names he had already counted, Count Disorderly kept on counting. He got 27. Then he counted again to be sure and got 22. He was all confused. The cook told him she couldn't get the party ready until she knew how many were coming. Poor Count Disorderly! He sat down and put his chin in his hands. Just then his brother, Count Orderly, arrived for a visit. "Hey, what's the matter? Aren't you excited about your birthday?" he asked. Count Disorderly responded, "Well, I was, but I can't figure out how many people are coming. I keep getting a different number when I count." Count Orderly took the list and told his brother they could count together. He whipped out a magic marker, and they started at the top of the list. Every time they counted a name, he checked it off. In this way, they counted each name on the guest list just once. There were 25 names! Count Disorderly skipped away to the cook to tell her.

Order Irrelevance

Count Disorderly had a big day planned, but he didn't dare get out of bed and go downstairs. Yesterday morning he'd counted the steps when he went down to breakfast and got 10. But when he came up to go to sleep he counted 11. If there were fewer steps going down than up, maybe he'd fall off the bottom today! So he sat and watched the sun come up. It was a beautiful day. The cook came and yelled up the stairs that his breakfast was getting cold. His friends came and yelled that they were ready to leave for a hike. But Count Disorderly wouldn't come down. Everyone left. Then Count Orderly showed up and ran right up the stairs to ask his brother Count Disorderly what the problem was. When he heard that his brother thought he'd fall off the staircase, he exclaimed, "It can't be! The stairs have the same number of steps whether you go up or down!" He dragged Count Disorderly out of bed and to the top of the stairs. Count Disorderly was scared but grateful to his brother for risking a fall. Count Orderly walked down the stairs and counted each one: "10!" Then he came back up, counting again, and got 10. "It's the same stairs, so it will be the same number of steps," said Count Orderly. Count Disorderly jumped up and down, thanked his brother many times, and then ran down the stairs to go outside and catch up with his friends on their hike.

These stories were written with the assistance of Cathy A. Mason.

ber to define equivalence and inequivalence, they typically have considerable difficulty on such tasks. After ensuring that a child has accurate numbering skills, it may be helpful to point out explicitly how counting can be used to determine same as, different from, and more than. This can be done in the contexts of games such as those described in Example 7.2. Games like lotto have been used successfully even with mentally handicapped children (Carrison & Werner, 1943; Descoeudres, 1928).

Basic Arithmetic Concepts

A fundamental understanding of arithmetic is not likely to develop without proficient counting skills and sufficient counting experiences. Without accurate and ample numbering experiences, a child will not learn the effects of adding an item to a set: Increments of one systematically change the cardinal designation of a set to the next number in the number sequence. Thus remedial efforts for arithmetic should not be undertaken until a child is proficient with basic counting skills such as enumeration, the cardinality rule, and even production. Particularly for special education children, it may be helpful to point out the effects of adding or taking away one in everyday situations. For example, at snack time, a teacher may give a child two cookies and ask how many adding one more to the two would be. Or after a child has eaten one cookie of the three, ask how many are left. Several games that involve keeping score of increments and decrements of one are illustrated in Example 7.3.

Number and Finger Patterns

By the time they enter school, children typically can subitize sets of up to four items (Bjonerud, 1960; Gelman, 1977). Some disadvantaged and many mentally handicapped children will not have mastered this basic skill (Baroody & Ginsburg, 1984). Subitizing sets of five and six, or even three or four, may actually depend on accurate numbering skill and ample counting experiences. Thus deficiencies in these areas should be remedied before the child is expected to master pattern recognition. Recognition of regular patterns can be cultivated by playing games that use a die.

At least for the numbers 1 to 5, many children spontaneously learn automatic finger patterns before they enter school (Siegler & Robinson, 1982; Siegler & Shrager, 1984). Such a skill cannot be taken for granted in special populations. Several activities for encouraging the learning of finger patterns are described in Example 7.4.

(*Text continues on p. 122*)

EXAMPLE 7.2

Games for Teaching Equivalence, Inequivalence, and Magnitude Concepts

LOTTO

Objective: Equivalence and inequivalence

Materials: (1) Boards for each player
(2) Squares with different numbers of dots

Instructions: Each player takes a board with, say, three number patterns (see figure below). On their turn, children try to find a square that has the same number of dots as one of their number patterns. The matching square is placed on the appropriate number pattern. The first player(s) to fill up their board (cover all their number patterns) wins. Once a turn is started every child can be given a chance to play. This eliminates the advantage of going first and allows for more than one winner.

(continued)

EXAMPLE 7.2 (continued)

DOMINOES SAME NUMBER

Objective: Equivalence and inequivalence

Materials: Set of dominoes

Instructions: This game is adapted from the domino game described by Carrison and Werner (1943) and Wynroth (1969/1980). Turn the dominoes face down. Players pick an equal number of dominoes. Player with the double two starts. First player to use up all his or her dominoes wins. Play with a regular set of dominoes is illustrated below. To encourage a greater reliance on counting, Wynroth (1969/1980) uses dominoes with an irregular arrangement of dots, which makes pattern recognition less likely.

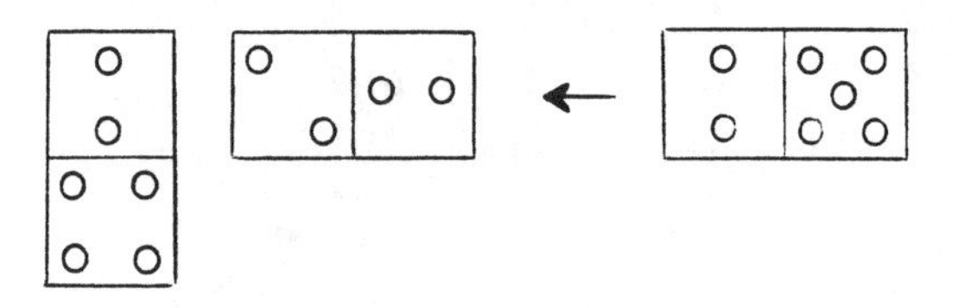

STAIRCASE

Objectives: (1) Number sequence represents increasingly larger quantities (introductory magnitude concept)
(2) The next number in the sequence is a unit or one larger (more advanced concept)

Materials: Interlocking blocks

Instructions: Help the child to build a staircase with interlocking cubes. Use alternate colored cubes to highlight the increment in units. As the child builds the staircase, point out that the first step has only one block and is not very big, the next step has two blocks and is a bit (one block) bigger, the following step has three blocks and is bigger yet (one block bigger than two), and so on. After the staircase is built (up to 5 or even 10 steps), have the child "walk up" the steps with his or her fingers, counting each step as it is touched. The staircase can also be built in conjunction with a number list. Again, point out that, as the child goes further along the number list, the numbers (steps) get bigger (that each successive number or step is one block bigger).

EXAMPLE 7.3

Games Involving the Addition and Subtraction of One

PENNY PITCH

Objective: Adding by one to 5

Materials: (1) Pennies or other small countable objects
(2) Plates (differently colored sheet of paper)

Instructions: The aim of the game is to pitch a winning number of pennies onto a plate. Each player chooses a different-colored plate. For beginners, set the winning number at "five." On their turn, players pitch a single penny. If a child is successful on his or her turn, say, "You had three pennies on the plate, and you got one more on. How much is three and one more? If a child is unable to figure out an answer any other way, add: "To see how much three and one more is, count the pennies on your plate." The first player to get to five wins. The difficulty of the game can be varied by changing the distance to the plate or the score needed to win.

THE COOKIE MUNCHER GAME

Objective: Subtracting by one

Materials: (1) Deck of (index) cards with 1 to 5 cookies (signal dots, circles, or pictures of cookies)
(2) Round countable objects

Instructions: The aim of the game is to collect 10 cookies (countable objects). On their turn, the players pick a card and can take one less cookies than indicated. Explain: "The cards tell us how many cookies you can take on your turn. However, the storekeeper Cookie Monster always eats one cookie from each order he serves." After a child has chosen the cards with 3 dots, for example, say: "You are supposed to pick up three cookies, but Cookie Monster eats one. How many are left for you to take?" If the child is correct, say, "You can take two cookies." If a child is unable to answer, have the child cover one dot with a finger and count the rest. For some children a more concrete demonstration may be necessary: After three cookies are put out and one is taken away by a Cookie Monster muppet, have the child count the remaining cookies. Summarize the event by saying, "There were three cookies, one was taken away, and that left two."

EXAMPLE 7.4

Activities for Learning Finger Patterns

FINGER PUPPETS

Objective: Automatic finger representation of numbers 1 to 10

Materials: Finger puppets made of paper and taped to slip on top of finger or stickers with faces to put on fingers.

Instructions: Show the child the right fingers to put up by attaching puppets or stickers to them.

HAND SHAPES

Objective: Automatic finger representation of numbers 1 to 10

Materials: (1) Chalkboard
(2) Chalk

Instructions: Help child put up the correct fingers for various numbers and trace them on the chalkboard. Then ask him to show you various numbers of fingers, and he can check his responses against the traced shape or by putting his hand against yours with correct shape.

These activities were suggested by Mary Loj.

EDUCATIONAL IMPLICATIONS: THE NATURE OF BASIC INSTRUCTION

Different Views—Different Implications

The logical-prerequisites and counting views have profoundly different educational implications. According to the first, it is pointless to focus initial training efforts directly on counting and number skills. Van Engen and Grouws (1975) note: "The notion that counting is the basic idea of arithmetic has been accepted and promoted for a long time by many people interested in elementary school mathematics. Counting is *not* the most basic idea of arithmetic! Ideas such as one-to-one correspondence and "more than" are much more fundamental and are, in fact, prerequisites to a meaningful development of counting" (pp. 252–53). Without the general psychological prerequisites, counting and number instruction is doomed to be meaningless. Therefore mathematics instruction must first promote the development of logical concepts and reasoning. According to the second view, initial instruction needs to focus directly on the development of specific counting skills and concepts and encourage their application. In brief, the issue is whether primary mathematics instruction should be taught formally by building on from more basic logical concepts or informally through counting.

THE NEW MATH. During the 19th century and most of the 20th century, mathematical instruction of the young has begun with counting (Brainerd, 1973). Dewey (1898) and Thorndike (1922), for example, concluded that counting should comprise the child's initial training in mathematics. Russell (1917) denounced this informal approach. He argued that the logical concept of classes should be taught first and that number should be taught later as the by-product of these ideas. Russell's "cardinal approach to early mathematics instruction" finally took hold in the form of the "New Math" (Brainerd, 1973).

The cardinal approach, or New Math, emphasizes teaching set theory. Figure 7.1 illustrates the first lessons in this approach. What concepts are the exercises on page 5 intended to cultivate? What is the objective of the exercises on page 6? What is the aim of the exercises on page 7? As Figure 7.1 illustrates, initial instruction focuses on cultivating concepts of classes (classification and class inclusion) and equivalence (one-to-one correspondence).

However, as noted in Chapter 2, such a formal approach is foreign to young children. Consider the case of Aaron, a bright and lively lad, who had just begun first grade. The year before, I had tracked his relatively quick development of informal addition. In a matter of months he had mastered adding with blocks and his fingers. Then he went on to invent mental computing procedures. Curious about his progress, I asked Aaron how he liked math this year. He shrugged his shoulders unenthusiastically. I asked what he was learning about in math.

AARON: [Without interest.] I'm not sure. We have to draw lines and junk like that.

INTERVIEWER: Oh, you're matching sets to see if they're equal.

AARON: I guess. [Then his whole demeanor was transformed by a surge of enthusiasm.] Do you know how much 1,000 plus 1,000 is? It's 2,000!

INTERVIEWER: Wow! Did you learn that in math class?

AARON: No, I'm just smart!

Because he did not seem to understand the point of the matching exercises, Aaron took little interest in this formal approach. Yet he was capable of

FIGURE 7.1: The First Pages of a Beginning Math Workbook

Page 7

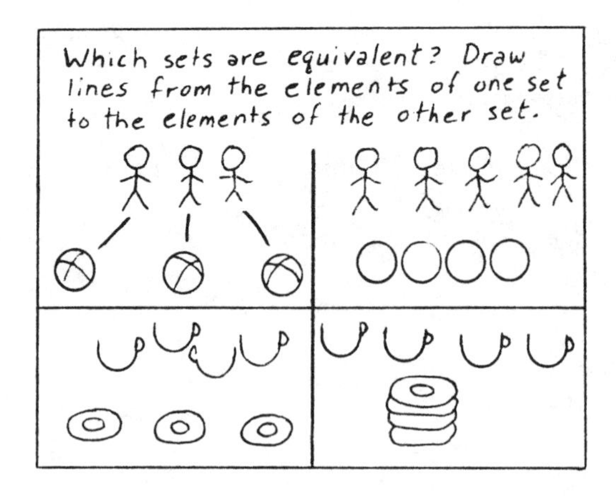

understanding basic arithmetic and extending a relationship that he learned with single-digit addends to a four-digit level. His informal observation was meaningful and exciting to Aaron.

PIAGETIAN INSTRUCTION. Some Piagetian educators argue that because early stages of intellectual development place limits on children's ability to understand number, initial mathematics instruction should be designed to encourage the development of operational thinking (e.g., Copeland, 1979). A number of curricula (e.g., Furth & Wachs, 1974; Maffel & Buckley, 1980; Sharp, 1969) have been designed with the broad aim of encouraging general thinking (logical) ability.

In the Piagetian view, it is pointless to teach number (counting and arithmetic) directly. One must first encourage the psychological prerequisites: an understanding of classes, relations, and one-to-one correspondence. This view is reflected by Gibb and Castaneda (1975) in a yearbook of the National Council of Teachers of Mathematics: "Classifying, [matching,] and ordering are three processes that underlie the concept of number. . . . Hence experience in classifying, comparing, and ordering provides the necessary background for the higher degree of abstraction required for number" (p. 98). Developing the meaning and names for numbers and counting should come only after much experience with classifying, ordering, and matching (Gibb & Castaneda, 1975).

However, there is little evidence to justify this Piagetian approach to initial primary instruction. Indeed, the evidence (Almy, 1971; Dodwell, 1960, 1962; Gonchar, 1975; Hood, 1962) seems to support the argument (Macnamara, 1975) that number is not dependent upon the development of formal classification and serial-ordering skills as described by Piaget. Moreover, an ability to compare sets by means of counting is not dependent upon one-to-one matching competence (e.g., Wang, Resnick, & Boozer, 1971). Children can learn much about counting, number, and arithmetic before they can conserve (Mpiangu & Gentile, 1970). Indeed, the need to posit stages of logical development has been seriously questioned (e.g., see Groen & Kieran, 1982). In brief, it has not been empirically demonstrated that success on "operational" tasks such as class inclusion, serial ordering, one-to-one matching, and number conservation is necessary for a basic understanding of number, counting, and arithmetic.

Nevertheless, the Piagetian position does have a number of important educational implications. For example, an elementary notion of "more than" is necessary for the development of a number concept and meaningful counting. Moreover, number does have *both* ordering and classification meanings, and counting does involve one-to-one correspondence. However, it may be that these concepts are attained by children in their *basic* forms earlier than Piaget thought and that number and counting only require an informal understanding of these concepts. Indeed, the development of a *more elaborate* and formal grasp of classification, more than, and one-to-one correspondence may actually depend on the development of counting and number.

Curriculum Implications

The goal of the New Math and Piagetian curricula, to help children think logically, is without question important. Reasoning about classes and relations should be an aspect of elementary mathematics curricula. Initial mathematics instruction should, however, take into account what is meaningful to young children. Some recommendations follow:

1. *Introduce mathematics informally rather than formally through set theory*. Formal definitions of number equivalence and so forth may be too abstract for young children. Counting provides a concrete and meaningful basis for understanding such key ideas as equivalence, inequivalence, and number conservation, especially with nonintuitive sets. Indeed, counting, for instance, may be more meaningful than matching in establishing the equivalence of sets, particularly those with more than five objects.
2. *Do not postpone counting experience and training*. Even preschoolers seem psychologically equipped to begin learning about number. Except for basic notions of "more," there is no need to delay counting instruction for general skills such as classifying, serial ordering, or matching. It is important to teach these skills for their own sake, but there is little reason to believe these skills are necessary for number and counting training. Furthermore, there is also no need to delay counting, number, and arithmetic training for nonconservers.
3. *Encourage the development of automatic pattern recognition and finger patterns*. Subitizing has sometimes been dismissed as a rotely learned skill—a skill that is more readily obtained than enumeration or a number concept (e.g., Strauss & Lehtinen, 1950). Recognition of number patterns plays an important role in the development of number and arithmetic. Children should be encouraged to master regular number patterns, such as those on dice. In addition, they need experience with nonregular arrays of one to five. By automatically recognizing various number patterns as instances of the same number, children can learn that number and equivalent sets are not defined by appearance. Finger patterns also play an important role in number development and, as we will see in Chapter 8, in arithmetic development. Young children should therefore be encouraged to count on their fingers and use finger patterns.

SUMMARY

Counting experience is central to children's slowly evolving understanding of number and mastery of number applications. Except to remedy very basic notions, such as "more," there is little reason to postpone counting and number training. From concrete counting and pattern-recognition experiences, children learn that changes in appearance and counting order do not affect cardinality but that adding or taking away elements does. Counting experience is important to extending intuitive notions of equivalence, inequivalence, and more than. Formal set theory and logical training are useful in their own right, but number training based on counting is initially more meaningful to children.

CHAPTER 8

Informal Arithmetic

Before mastering the basic number combinations, what kinds of procedures do children use to figure out sums, differences, and products for single-digit problems? What accounts for the development of informal-arithmetic procedures? Why are some problems more difficult for children to calculate than others? How do children naturally try to minimize the difficulties of calculation? What kinds of problems do children often encounter with informal-arithmetic calculation? How can these difficulties be remedied?

THE BASES FOR INFORMAL ADDITION AND SUBTRACTION

Problems with Mental Computing: The Case of Aaron

At the beginning of kindergarten, 5-year-old Aaron could quickly generate the sums for plus-one ($N + 1$) problems, such as $3 + 1 = \_$ and $5 + 1 = \_$. For other problems, including one-plus-a-number ($1 + N$) problems such as $1 + 3$ and $1 + 5$, Aaron had to use concrete objects to compute the sums. Take, for example, his responses in our fourth interview in November:

INTERVIEWER: 1 + 7.

AARON: [Pause. Then counts to himself "1, 2, 3, 4, 5, 6, 7."] I have to do it with the blocks. [Puts out one block, produces seven more, counts all the blocks, and announces the correct sum.]

INTERVIEWER: 2 + 3.

AARON: [Quickly counts] 1, 2; 3 [pause]. I'm almost there, but I ran out of thinking. [Puts out two blocks, then three more, and counts all the blocks to determine the sum.]

INTERVIEWER: 2 + 4.

AARON: 1, 2, 3, 4, 5, 6, 7, 8, 9, 10. I don't know . . . [Again uses blocks to compute the sum.]

INTERVIEWER: $1 + 3$.
AARON: 1, 2, 3, 4, 5 [quietly]. [Interviewer: Close.] Four.

Aaron's mental-addition ability grew gradually. With $1 + N$ problems, he initially counted from one (e.g., $1 + 3$: "1, 2, 3, 4"). By the spring, he automatically solved $1 + N$ problems by announcing the number after N in the number sequence (e.g., $1 + 3$: "4"). By the end of the school year, Aaron elaborated on this procedure to compute the sums of non-one problems: start with the cardinal term of the larger addend and count on (e.g., $2 + 6$: "6; 7, 8").

Aaron received almost no formal arithmetic instruction. What accounts for his informal arithmetic strengths and his progress during the year? Why did he calculate the sums of $1 + N$ problems when he could readily determine the sums of $N + 1$? In contrast to his proficiency in calculating sums with concrete aids, his initial mental efforts were not successful. What accounts for the difficulty in developing mental computing procedures for $1 + N$ and non-one problems? During the year, Aaron spontaneously invented new computing procedures. What motivated this development?

The Foundation: Counting

As we saw in Chapter 7, a fundamental understanding of arithmetic evolves from children's early counting experiences well before they begin school. Informal concepts of addition (as adding *more*) and subtraction (as taking away something) guide children's efforts to construct informal-arithmetic procedures. To add one more to three, say, many children first count up to three and then simply count once more ("1, 2, 3; *4*"). Children may even try to tackle more difficult problems in the same manner. Consider Aaron's first attempts to mentally compute the sums for $1 + N$ and non-one ($M + N$) problems. Because he viewed addition as an incrementing process, his initial efforts—though not successful—were on the right track. For $2 + 3$, for example, he appeared to know that the sum had to be larger than two. Therefore, he quickly counted up to two, counted once more (but was unsure how to proceed): "1, 2; 3 . . . I'm almost there but. . . . "

Proficiency with counting skills enables children to mentally solve one problems quite early. Children rather quickly discover that number-after relationships apply to $N + 1$ problems and that number-before relationships apply to $N - 1$ problems. In fact, many preschoolers can use their mental representation of number sequence to solve simple one ($N + 1$ and $N - 1$) problems, such as "three candies and one more" or "five dolls take away one" (e.g., Baroody, 1984a; Court, 1920; Fuson & Hall, 1983;

Gelman, 1972, 1977; Ginsburg, 1982; Groen & Resnick, 1977; Ilg & Ames, 1951; Resnick, 1983; Resnick & Ford, 1981; Starkey & Gelman, 1982). For the addition problem above, for example, a child enters the number sequence at a point specified by the first term or augend (three) and gives as an answer the number *after* this in the number sequence: "Four." For the subtraction problem cited, a child enters the number sequence at the point specified by the minuend, or larger number (five), and gives as an answer the number *before* this in the sequence: "Four." Because using their mental representation of the number sequence to determine number-after and number-before responses is so automatic, many preschoolers can mentally and quickly produce the answers to simple one problems.

The Relative Difficulty of $1 + N$ Problems

Why could Aaron solve $N + 1$ but not $1 + N$? Children's informal concept of addition initially may make $N + 1$ problems easier to solve than $1 + N$ problems. For example, because he viewed addition as an incrementing process, Aaron interpreted the problem $3 + 1 = \_$ as three and *one more*, which is readily solvable by counting ("1, 2, 3; 4") or using number-after relationships ("3, 4"). In contrast, he interpreted $1 + 3 = \_$ as one and *three more*, which is *not* readily solvable by these methods. In other words, because they view addition as an incrementing process, young children may tend to view $N + 1 = \_$ and $1 + N = \_$ as *different* problems and *non-equivalent* in sum. Thus, they may not realize that their number-after method for quickly answering $N + 1$ is also applicable to $1 + N$.

At some point, children discover that number-after relationships apply to $1 + N$ as well as $N + 1$. Jenny, a kindergartner, described this important discovery (Baroody & Ginsburg, 1982a). While playing a math game, the girl sitting next to Jenny drew the problem $1 + 6 = \_$. Obviously perplexed and unsure how to solve the problem, the girl remained silent. After a moment, Jenny leaned over and whispered to the girl, "Oh that's easy! Whenever you see one, it's just the next number!" Unlike her peer, Jenny had abstracted a general number-after rule for one problems: "The sum of $N + 1$ *or* $1 + N$ is the number after N in the number sequence." With this general rule, Jenny could use her mental representation of the number sequence to efficiently answer $1 + N$ as well as $N + 1$ problems.

The development of a general number-after rule for one problems may be an important first step to more flexible computing generally. For example, Aaron first learned that he could safely disregard addend order with one problems. Some weeks later he began disregarding the order of non-one problems and computing the sums of $M + N$ by counting on from the larger term (e.g., $2 + 6$: "6; 7, 8"). Moreover, children only gradually come to view addition as the *union* or joining of two sets. In this view, the order of numbers is unimportant: $3 + 2 = 2 + 3$. In other words, the union of a set of three things and a set of two things has the same outcome as the joining of two things and three things. This joining view of addition is more abstract than the incrementing view that is familiar to young children. The realization that addend order does not affect the outcome of one problems may be an important first step toward this deeper understanding of addition (Resnick, 1983).

INFORMAL ADDITION

Concrete Procedures

Initially, children use concrete objects to figure out sums. Because of their ready availability, fingers are commonly used with sums up to 10. Developmentally, the most basic strategy is concrete counting all (CC), illustrated in column 1 of Figure 8.1. Blocks (or other countable objects, such as fingers) are counted out one by one to represent an addend; the process is repeated for the other addend. Then all the objects are counted to determine the sum.

INVENTED SHORTCUTS. Children spontaneously invent shortcuts for the cumbersome CC procedure. A favorite shortcut is the "finger-pattern" strategy illustrated in column 2 of Figure 8.1 (Baroody, in press). Note that, for this strategy, each addend is represented by a finger pattern. Thus, the laborious process of counting out fingers one at a time to represent each addend is circumvented. With the finger-pattern strategy, the child has to count only once (to determine the sum). The "pattern-recognition" strategy (Siegler & Robinson, 1982; Siegler & Shrager, 1984), illustrated in column 3 of Figure 8.1, is even more economical. This strategy involves creating finger patterns for each addend and then immediately recognizing the sum—perhaps visually (through subitizing) or kinesthetically. For $1 + 2 =$ __, for instance, a child can immediately see and grab one block, do the same with two blocks, and immediately recognize that the total is "three." For $4 + 5 =$ __, for example, a child can use finger patterns to represent each addend, sense that all but one of the ten fingers are raised, and, without counting, respond, "Nine."

To save labor, even mentally handicapped children will spontaneously invent CC shortcuts. Mildly and moderately handicapped children who relied on or had to be taught a CC procedure were observed over a period of 21 weeks. Without instruction or encouragement, many of the study

FIGURE 8.1: Concrete Addition Procedure to Solve 4 + 2 = __

Step	Concrete Counting-All	Finger-Pattern Strategy	Pattern-Recognition Strategy
1	Count out objects to represent the first addend: "1, 2, 3, 4" □□□□	Put up a finger pattern to represent the first addend: "4" 	Put up a finger pattern to represent the first addend: "4" 
2	Count out objects to represent the second addend: "1, 2" □□□□ □□	Put up a finger pattern to represent the second addend: "2" 	Put up a finger pattern to represent the second addend: "2" 
3	Count all the objects to determine the sum: "1, 2, 3, 4 5, 6" □□□□ □□	Count all the fingers to determine the sum: "1, 2, 3, 4 5, 6" 	Recognize the results: "Oh, 6" 

participants began using a finger-pattern strategy, and a few used a pattern-recognition strategy for problems with very small addends. Thus it seems that a tendency to invent calculational shortcuts is common to children with a very wide range of mental ability.

SELF-MONITORING, INVENTION, AND FLEXIBILITY. By monitoring their efforts, children can adapt existing procedures to meet new demands and thus invent new procedures. Consider the case of Mike, a 20-year-old with an IQ of 46. Mike encountered a situation common to children: His concrete procedure served him well as long as the numbers remained small. For problems with addends of five or less, Mike used a finger-pattern procedure (e.g., for 3 + 5, he put up finger patterns of three and five on separate hands and then counted all eight fingers). However, this procedure cannot be easily used with problems, such as 2 + 8 and 6 + 3, where each addend cannot be as readily represented on one hand.

Apparently aware of the limitations of his finger-pattern procedure with larger problems, Mike modified the procedure. When presented with 2 + 8 and 6 + 3, he first put out a finger pattern of only the smaller addend (e.g., two fingers for 2 + 8). With a cardinal model of the smaller addend already in place, his next step was to start with one and count up to the cardinal designation of the larger term (e.g., "1, 2, 3, 4, 5, 6, 7, 8" for 2 + 8). Then Mike simply continued this count as he pointed to each element in the cardinal model created ahead of time in Step 1 (e.g., for 2 + 8: "9 [points to the first of two fingers up], 10 [points to the second finger] – 10)." Mike's preplanned maneuver to represent only the smaller addend and then use this model to count on from the larger addend enabled him to cope with larger problems.

Mike's relatively sophisticated concrete ("ready-model") procedure appears in other ingenious forms. Some children use cardinal models already present in the classroom to count on. Some children use the points on numerals: 2.3.4 (e.g., 2 + 4: "1, 2; 3 [point to the top-left point of the four], 4 [point to the top-right point], 5 [point to the middle point], 6 [point to the bottom point] – 6"). While doing their arithmetic seatwork, some children look at the clock a great deal (for reasons other than checking the time till recess). The clock provides a ready-made model for counting on (e.g., 2 + 4: "1, 2; [look at 1 o'clock] 3, [look at 2 o'clock] 4, [look at 3 o'clock] 5, [look at 4 o'clock] 6 – 6"). Some children may even mentally create a model in order to keep track. For example, for 2 + 4 a child might imagine four dots in the pattern of a box and count "3, 4, 5, 6" as she "pointed" or "looked" at the imaginary dots. A ready-model procedure may be a springboard for inventing efficient mental computing procedures (Fuson, 1982).

Furthermore, monitoring enables children to choose intelligently among informal addition procedures (Siegler & Robinson, 1982). For example, Kathy, a 15-year-old with an IQ of 40, used a finger-pattern strategy for problems such as $2 + 3$, $4 + 2$, and $5 + 4$. With problems such as $2 + 8$ and $6 + 3$, she immediately recognized the limitations of her finger strategy and, without prompting, switched to computing with blocks (fell back on a CC procedure). Moreover, when presented with $1 + 2$ and $3 + 1$ (combinations that previous testing indicated that she knew), Kathy did not mechanically continue to use concrete procedures (perseverate) but responded automatically. Thus it appears that even some children with significant mental handicaps monitor their arithmetic work and exhibit flexibility in choosing computing procedures.

Mental Procedures

In time, children spontaneously abandon concrete procedures and invent mental counting procedures for computing sums (Groen & Resnick, 1977). The most basic mental addition procedure is counting all starting with the first addend (CAF) (e.g., $2 + 4$: "1, 2; 3 [is one more], 4 [is two more], 5 [is three more], 6 [is *four* more]—6") (Baroody, 1984a, in press; Baroody & Gannon, 1984). CAF is a fairly sophisticated invention because it does not directly model the whole concrete counting-all process, and it entails enumerating the addend *as* the child counts on from the augend (a simultaneous keeping-track process) (Baroody & Ginsburg, in press; Carpenter & Moser, 1983).

KEEPING TRACK. This keeping-track process makes mental computing of non-one problems more difficult than one problems. To compute $N + 1$ or $1 + N$, a child merely has to count up to N and then give the next number in the count sequence (e.g., $1 + 4$: "1, 2, 3, *4*; [and one more is] 5"). With non-one problems, a child must continue the count beyond N a specified number of times. Accurate mental computation of non-one problems requires a preplanned means of keeping track. This is what prevented Aaron from successfully computing non-one problems early in the school year. In time, Aaron spontaneously devised keeping-track methods.

Especially at first, children use concrete objects to keep track. Fingers are a favorite means of keeping track (e.g., $2 + 4$: "1, 2; 3 [one finger extended is *one* more], 4 [two fingers extended is *two* more], 5 [three fingers extended is *three* more], 6 [four fingers extended is *four* more]—6"). If a child can automatically recognize finger patterns, the keeping-track process requires little attention and thus can be executed quite efficiently. For instance, for $2 + 4$, once the counting-on process is started, a child simply counts until he or she "feels" the fourth finger go up. Knowing the finger pattern for four tells the child when to stop.

In time, children go from counting objects to counting less concrete things to keep track (Steffe et al., 1983). Indeed, children use a variety of means to keep track (e.g., see Fuson, 1982). Some children tap their finger or pencil as they count on. They may even exploit patterns, such as drum-drum, drum-drum is four (e.g., 2 + 4: "1, 2; 3, 4 [drum-drum]; 5, 6 [drum-drum] – 6"). Children may also keep track with a second verbal or subvocal count: a double count (e.g., 2 + 4: "1, 2; 3 is *one* more, 4 is *two* more, 5 is *three* more, 6 is *four* more"). When children are very familiar with the number sequence, this double-count process can become highly automatic and accomplished mentally.

INVENTED SHORTCUTS. Counting on from the first addend (COF) shortcuts the CAF procedure by starting with the cardinal term of the first addend (e.g., 2 + 4: "2; 3 [+ 1], 4 [+ 2], 5 [+ 3] 6 [+ 4] – 6"), but it does *not* reduce the number of steps in the keeping-track process (Baroody & Gannon, 1984). Both CAF and COF entail a four-step keeping-track process. Because it does not save much effort, children only rarely invent and use COF (Baroody, in press; Baroody & Ginsburg, in press).

The cognitively demanding keeping-track process can be minimized by starting with the larger term. One strategy that accomplishes this end is counting all starting with larger term (CAL) (Baroody, 1984a). CAL entails beginning with "one," counting up to the larger cardinal term, and then counting on from there while the smaller term is enumerated (e.g., 2 + 4: "1, 2, 3, 4; 5 [+ 1], 6 [+ 2] – 6"). Note that, by starting with the larger term, the keeping-track process for 2 + 4 is reduced from four to only two steps. On the other hand, note that a CAL procedure requires an additional process not required by CAF: deciding which addend is larger. Children adopt CAL because deciding which addend is larger has become automatic and requires negligible effort in comparison to a keeping-track process.

Counting-on from the larger term (COL) shortcuts CAL by starting with the cardinal designation of larger term and hence is the most economical informal mental addition procedure (e.g., 2 + 4: "4, 5 [+ 1], 6 [+ 2] – 6"). During the course of their computational efforts, children may realize that counting out the larger addend is redundant with simply stating the cardinal term of the larger addend (Fuson, 1982; Resnick & Neches, 1984). As a result, they adopt the shortcut of starting with the cardinal term of the larger addend instead of counting from one. Thus CAL is abandoned in favor of COL because it saves even more labor.

SELF-MONITORING, INVENTION, AND FLEXIBILITY. As with concrete procedures, self-monitoring leads children to invent labor-saving mental procedures spontaneously by helping them sense when existing methods need to be readjusted. Consider the example of Casey, a kindergartner (Baroody & Gannon, 1984). During our first session together, Casey relied exclusively on CAF. During our second meeting, Casey was presented with 3 + 6 typed on an index card. He began as he usually had, counting out the first addend as he tapped the card: "1, 2, 3." This time, however, he stopped and commented, "I'll count to 6, I guess. 1, 2, 3, 4, 5, 6 [pause], 7, 8, 9." Apparently, Casey must have anticipated the difficulty of a six-step keeping-track process if he continued with his CAF procedure. To save labor, he adjusted his approach. He began by counting out the larger addend and thus invented a new, more easily executed CAL procedure.[1] As with concrete computing procedures, it appears that children are self-motivated to minimize cognitive effort.

Self-monitoring also permits children to choose flexibly among mental procedures. Casey, like many children (e.g., Carpenter & Moser, 1984), did not use his newly invented and more advanced procedure consistently. Though he used CAL with problems like 2 + 6 = __, 2 + 8 = __, and 3 + 6 = __ that entailed a demanding keeping-track process, he continued to use CAF with problems like 2 + 4 = __ that required a less involved keeping-track process.

INFORMAL SUBTRACTION

Concrete Procedures

For problems with subtrahends (smaller numbers) greater than one, children—at first—usually use concrete representations to model directly their informal concept of subtraction as "take away" (e.g., Carpenter & Moser, 1982). Their separating-from procedure entails (a) representing the

1. It would seem that children who invent addition procedures that disregard addend order (CAL or COL) would understand commutativity (e.g., Groen & Resnick, 1977). However, it appears that Casey did not realize that 3 + 6 and 6 + 3 were equivalent: produced the same sum. In follow-up interviews, Casey indicated that 6 + 4 and 4 + 6, for example, would add up to something different (Baroody & Gannon, 1984). Children may add numbers in either order because they believe they will get a *correct* (though not necessarily the *same*) answer. It appears that nonconceptual factors, such as a drive to save mental labor, as well as conceptual knowledge help to account for the development of informal computing procedures.

minuend (larger number), (b) removing a number of items equal to the subtrahend, and (c) counting the remaining items to determine the answer. For example, 5 – 2 would involve counting out five fingers or objects (making five marks), counting and removing two of the items (crossing out two of the marks), and finally counting the remaining items (marks): "Three" (see Figure 8.2).

Mental Procedures

COUNTING DOWN—A NATURAL EXTENSION OF EXISTING KNOWLEDGE. As with addition, when ready, children will abandon concrete procedures in favor of mental procedures. A common mental procedure is counting down, which like separating from, follows from a take-away interpretation of subtraction (Carpenter & Moser, 1982). Counting down involves stating the minuend, counting backward a number of times equal to the subtrahend, and announcing as the answer the last number counted (e.g., 5 – 2: start with 5, 4 [that's *one* taken away], 3 [that's *two* taken away]—the answer is "3"). Though counting down is a natural extension of children's mental number-before-*N* procedure for figuring out *N* – 1 differences, it is cognitively more complicated. To figure out problems involving *N* – 1, a child merely had to know what number came before another in the number-word sequence. With larger subtrahends, the child must be able to count backward from a specified point

FIGURE 8.2: Separating-From Subtraction Strategy Using 5 - 2 as an Example

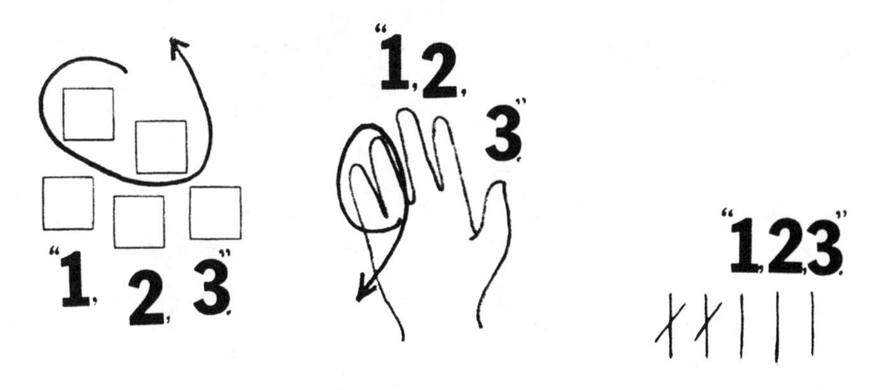

Five blocks, fingers, or marks are produced; two are removed, and the remainder are counted.

within the number word sequence a specified number of steps. Thus counting down entails a keeping-track process, which must be executed *while* the child counts backward (see Figure 8.3).

A DEMANDING PROCEDURE. The counting-down procedure for subtraction is also more difficult than children's informal mental addition methods (Baroody, 1984c). With mental addition procedures, *both* the sum count and the keeping-track process proceed in a *forward* direction. In contrast, counting down requires counting backward, which is more difficult than counting forward for young children (Fuson et al., 1982; Ginsburg & Baroody, 1983). Moreover, this relatively difficult backward count must be executed *while* a keeping-track process is executed in the forward or *opposite* direction! The imposing nature of this procedure may help to explain an alarmingly common refrain from primary schoolchildren: "I hate take away; it's so much harder than adding" (Starkey & Gelman, 1982).

The difficulty of the counting-down procedure is related to problem size. The size of the subtrahend is a key factor. In the case of 9 – 2, the keeping-track process is a relatively manageable two steps ("9; 8 [–1], 7 [–2] – 7"). In the case of 9 – 7, however, the keeping-track process is a very difficult seven steps ("9; 8 [–1], 7 [–2], 6 [–3], 5 [–4], 4 [–5], 3 [–6], 2 [–7] – 2"). For 19 – 17, the keeping-track process becomes a practically

FIGURE 8.3: Two Keeping-Track Procedures to Solve 5 - 2 = __

A. Using fingers to keep track of how many times to count backward (the subtrahend count):

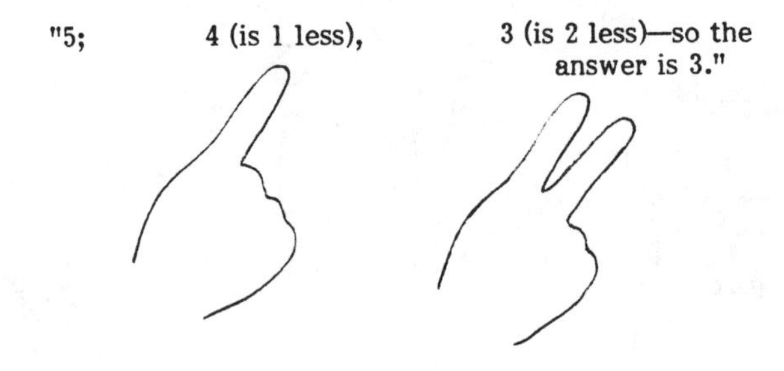

B. Using a double count to keep track of how many times to count backward (the subtrahend count): "Five, four is one taken away, three is two taken away—so the answer is three."

impossible seventeen steps ("19; 18 [−1], 17 [−2], . . . , 3 [−16], 2 [17] − 2"). Moreover, the size of the minuend can contribute to a child's difficulty. Counting backward from 20 is often more difficult for primary schoolchildren than counting backward from 10 (Fuson et al., 1982; Ginsburg & Baroody, 1983). Thus, as children are assigned problems with minuends in the teens (and beyond), the ones who rely on counting down are forced to use a less familiar and less automatic backward-counting sequence.

THE DEVELOPMENT OF PROCEDURE FLEXIBILITY. As their subtraction work involves increasingly larger numbers, it behooves children to learn or discover other subtraction procedures. Indeed, some evidence (Woods, Resnick, & Groen, 1975) indicates that many children first use a counting-down procedure and later invent a counting-up procedure. Counting up models a "missing-addend" approach to subtraction (Carpenter & Moser, 1982). It involves starting with the subtrahend and counting forward until the minuend is reached, while keeping track of the number of steps in this forward count (e.g., 19 − 17: "17, 18 [is one], 19 [is two]—so that answer is two"). Although counting up does not model a child's informal notion of subtraction as take away, under some circumstances it is cognitively easier to execute than counting down. When the subtrahend is relatively large, as in the case of 19 − 17, counting up greatly reduces the demanding double count or any other keeping-track process (to two steps, as opposed to seventeen steps with counting down). When the minuend and subtrahend are relatively close, as in 9 − 7, counting up also minimizes the keeping-track process (two steps, as opposed to seven steps with counting down). However, when the subtrahend is small and the minuend and subtrahend are relatively far apart (e.g., 9 − 2), counting down retains the edge in terms of ease of execution (a keeping-track process of two steps, as opposed to seven steps with counting up). By the third grade, many children discover this pattern and choose the more economical procedure (Woods et al., 1975).

INFORMAL MULTIPLICATION

Building Blocks

By the time many children are introduced to multiplication, they have acquired a solid basis for understanding multiplication and calculating products. Multiplication can be defined as the repeated addition of like terms (e.g., $5 \times 3 = 5 + 5 + 5$). Because it builds on familiar addition experiences, children readily assimilate multiplication. In the primary years, children have ample exposure to adding two or even more than two like sets, such as those depicted below:

••••   • •   • •   ••••

••••   • •   • •   ••••

••••

They have learned a variety of procedures to determine the total for, say, three groups of four. They may use skip counting ("4, 8, 12"), informal computing (e.g., "4 + 4 is 4; 5, 6, 7, 8 and 8 + 4 is 8; 9, 10, 11, 12"), known combinations (e.g., "4 + 4 is 8 and 8 + 4 is 12"), or some combination of these approaches (e.g., skip counting and computing: "4, 8; 9, 10, 11, 12").

Mental Procedures

Children initially rely on informal counting procedures to compute products (e.g., Kouba, 1986). As one girl explained, for $3 \times 3 = 9$, she had to "count three, three times" (Allardice, 1978). Most children just learning to multiply have the counting and keeping-track skills to calculate products mentally. Consider one child's explanation for computing 3×3: "Well, I just say the first number three out loud and then I say the next two numbers to myself, and then I say the next number out loud. So it's *three* [whispers 4, 5], *six*, [whispers 7, 8] *nine* . . . " (Allardice, 1978, p. 4). In effect, the child's informal mental procedure for multiplication involved three counts, including *two* keeping-track processes: (a) generate the number sequence, (b) keep track of every third number, and (c) keep track of how many groups of three to determine when to stop generating the number sequence. An even more basic mental procedure involves starting the number-sequence count at one (instead of the cardinal value of the first term or multiplicand). This basic procedure and the component processes are outlined in Table 8.1.

To make mental computing more manageable, children often create a set to represent the multiplicand. For 4×3, for instance, the child can use a finger pattern to represent the 4 and then just count the pattern three times. By using the finger pattern, the child eliminates the need to keep track of every fourth count. Previous informal experience may help children in other ways. Specifically, children quickly realize that skip counting

TABLE 8.1: Basic Mental Computing Procedure for Multiplication Using 4 x 3 as an Example

(a)	Generate successive numbers from the number sequence.	1 2 3 4; 5 6 7 8; 9 10 11 12
(b)	Keep track of every fourth number counted.	1 2 3 4 1 2 3 4 1 2 3 4
(c)	Keep track of the number of groups of four.	1 2 3
(d)	Stop generating the number sequence after completing the third group of four and announce the last number counted as the answer.	12

can be used in the service of multiplication (e.g., 5×3: "5, 10, 15"). Skip counting is a common procedure for figuring out products. Children also draw upon their existing formal knowledge to deal with multiplication. Children frequently use their knowledge of the addition doubles (e.g., "$4 + 4 = 8$") to determine products with a multiplier of two (e.g., 4×2) or to reason out larger problems (e.g., 4×3: "4 + 4 is 8 and four more is 9, 10, 11, 12" or 4×4: "4 + 4 is 8, that's two fours, 8 + 8 is 16, that's four fours").

INVENTED SHORTCUTS. As with informal addition and subtraction, children spontaneously find ways to shortcut informal calculation of products. Even children with learning difficulties can see ways of using their existing knowledge to shortcut computational effort. Consider the case of Adam, who was taught a concrete procedure for multiplying (e.g., 4×3: put out three groups of four blocks and count all the blocks). Almost immediately, this learning-disabled boy began to shortcut the concrete procedure. For instance, for 6×3 he put out three fingers and, instead of placing six blocks next to each finger, lined up six blocks for the first finger only. He counted the row of blocks ("1, 2, . . . , 6") and then counted the spaces where the blocks for the second and third row would have been placed ("7, 8, . . . , 12," "13, 14, . . . , 18"). He soon began to use mental procedures. For 4×3, Adam used a known addition fact ($4 + 4 = 8$) in combination with counting on (8, 9, 10, 11, *12*). For $5 \times N$, Adam very quickly realized that he could use skip counting (count by fives) to generate the answer.

EDUCATIONAL IMPLICATIONS: INFORMAL ARITHMETIC DIFFICULTIES AND REMEDIES

Plus and Minus One

Nearly all children entering school will have had sufficient informal experience to understand that addition is an incrementing process and subtraction is a decrementing process. Indeed, Starkey and Gelman (1982) found that nearly all of their 4-year-old subjects, if concrete objects were available, could solve $N + 1$ problems to $5 + 1$ by some means. Moreover, many 4-year-olds and the vast majority of 5-year-olds could do likewise with $1 + N$ problems to $1 + 5$. Moreover, by the time they begin school, most children have sufficient facility with number-after and number-before relationships to determine mentally and quickly the sums of $N + 1$ up to at least $5 + 1$ and the difference of $N - 1$ up to at least $5 - 1$ (Fuson et al., 1982). By the time they enter second grade, most children should be able to generate automatically the sums to $N + 1$ *or* $1 + N$ up to 10 and differences to $N - 1$ up to $10 - 1$. However, the incidental learning of basic informal arithmetic concepts and prerequisite counting skills cannot be taken for granted in special populations, especially with mentally retarded children.

1. *Ensure number-after (number-before) skill before mental addition (subtraction with one).* If children cannot automatically determine number-after (number-before) relationships, they will not be able to determine mentally and efficiently $N + 1$ sums ($N - 1$ differences). In such cases, remedial efforts should first focus on the prerequisite counting skill: the efficient use of the mental number list to determine N-after (N-before) relationships (Baroody, 1984b). During such remediation, children should be encouraged to use the concrete addition (subtraction) discussed in the following sections.

2. *Encourage discovery of a general number-after rule.* If a child can automatically solve $N + 1$ but not $1 + N$, create opportunities for the child to discover a general number-after rule. One strategy is to give the child a series of word problems in which $N+1=\_\_$ is followed by its $1+N=\_\_$ counterpart (or vice versa). For example, have the child solve Problem A below and then Problem B. After the child has computed Problem B, it may be helpful in some cases to ask the child if there were any similarities or differences with the previous problem.

   A. Sol had three cookies. He got one more from his mother. How many cookies did Sol have all together?

   B. Tammy had one cookie. She got three more from her mother. How many cookies did Tammy have all together?

For further practice, play games that involve special dice: a "one die" (a die with one dot on all six sides) and a die with just two, three, and four dots on the sides. The random process of rolling the dice will expose children to $2+1$ and $1+2$, $3+1$ and $1+3$, and $4+1$ and $1+4$ combinations, giving them many opportunities to see that $1+N$ problems add up to the same thing as their $N+1$ counterparts.

Addition

1. *Ensure readiness to learn informal addition procedures.* Though many children learn a concrete procedure for computing sums before they enter school (e.g., Baroody, in press; Carpenter & Moser, 1984; Lindvall & Ibarra, 1979), the development of a CC procedure cannot be taken for granted with all preschoolers—particularly disadvantaged children (Ginsburg & Russell, 1981) and mentally handicapped children. Though some children require only one or two demonstrations to learn a CC procedure, some children have difficulty mastering the procedure even after numerous demonstrations (Baroody, in press). Difficulty in mastering CC appears to be associated with a weakness in prearithmetic skills, such as comparing number neighbors. Moreover, deficiencies in basic counting skills will prevent children from inventing more efficient computing procedures. For example, an inability to automatically compare number neighbors will forestall adopting procedures that disregard addend order (CAL and COL). If children do not know number-after relationships automatically, it will be difficult for them to learn counting-on procedures (COF and COL). In such cases, readiness skills need to be mastered first.

2. *Use an incrementing model to introduce addition meaningfully.* Addition instruction is often introduced to children as the union of two sets. Children are taught a CC procedure, which more directly models a union-of-two sets view rather than an incrementing view of addition. For some children—especially lower-functioning children—it may be more useful to introduce addition in a manner that is consistent with a psychologically more basic incrementing model. That is, instruction can start with problems in which one or two is added to an existing set.

3. *Start with small-number problems; introduce larger problems slowly and carefully.* Initial addition training should involve small-number addends (1 to 5) that are easily handled by concrete procedures. This permits children to master a CC procedure and to invent shortcuts for CC to build a secure basis for further development. Introduce larger problems after a child can securely use concrete procedures with small numbers. This may provide an incentive to develop even more powerful mental computing procedures. However, some children may simply be overwhelmed by more

difficult problems. Therefore the introduction of larger problems should be monitored carefully.

4. *Allow for an extended period of computing and discovery.* If given the opportunity to use objects to compute sums, children typically will—at their own pace—invent mental procedures. For example, Groen and Resnick (1977) taught preschoolers a concrete addition procedure. After an extended period of practice and without counting-on instruction, about half the children began using COL. Many learning-disabled and some mentally retarded children will, without direct instruction, abandon concrete procedures and invent mental procedures like COL.

However, some children—especially disadvantaged children—may rely on concrete procedures for a long time. It is important to give children the opportunity to construct their own more advanced procedures, because self-invented procedures may be more meaningful to a child. In some cases, it may be helpful to model a COL procedure for the child. Such a demonstration will probably be more effective after children have had ample computing experience with objects. Direct verbal instruction is the least desirable method, because it is difficult to describe a mental procedure such as COL and explanations may serve only to confuse children (Resnick & Neches, 1984). In any case, I would not advise teaching counting on much before the middle of the first grade.

To facilitate the learning of mental procedures, a teacher should create ample opportunities for self-discovery. An interesting way to achieve this objective is to play games with dice. As children become familiar with the dice patterns, they usually find their own shortcut for determining the sum of a throw. For example, a child may roll a four (∷) and a two (⋅ .). If the child automatically recognizes the first pattern ("Oh, four"), he or she may not bother to start with one and count all the dots. The child may simply count on from four: "4; 5 [pointing to the first dot of the second die], 6 [pointing to the second dot]—6."

5. *Remedial efforts may have to focus on explicitly teaching a keeping-track process.* Sometimes children—especially disadvantaged and special education children—get stuck at the concrete level and seem unable to acquire mental procedures (notably COL) (Baroody, Berent, & Packman, 1982; Bley & Thornton, 1981; Cruickshank, 1948). In extreme cases, direct intervention may be necessary for learning mental procedures. Specifically, some children may need guidance to learn a keeping-track process. To teach a keeping-track process start with $N + 2$ or $2 + N$ and $N + 3$ or $3 + N$ problems. Introduce the idea of keeping track by teaching the child the procedure outlined previously in the case of Mike and summarized below:

A. Have the child focus on the smaller addend and create a set with fingers or blocks (e.g., for 4 + 2, produce two fingers).

B. Then have the child count up to the cardinal value of the larger addend ("1, 2, 3, *4*").

C. Then continue the count on the smaller set of objects previously produced ("5 is *one* more; 6 is *two* more—six").

6. *Encourage the learning and use of efficient keeping-track methods.* Automatic recognition of patterns can facilitate keeping track. Children should be encouraged to use and share their methods of keeping track. Children who have not mastered pattern-recognition skills, such as the finger patterns to 10, should be encouraged to master these skills.

The efficiency of children's mental computing often suffers when the second addend of problems is greater than five. Because it is difficult to keep track accurately, children often miscompute such problems. If they do resort to using fingers, computing may be accurate but very slow. Children must put their pencil down, count, pick up their pencil, and record the sum. Fuson (1985) suggests using Chisenbop finger patterns so that the numbers one to nine can be represented on the nonwriting hand, leaving the writing hand free to write with (see Figure 8.4). This device allows children to keep track of larger addends in a natural way—by matching successive number words to finger patterns.

Some slower children tend to forget the value of the second addend and thus lose track of when to stop the counting-on process. In such cases, Fuson (personal communication, July 28, 1986) has found it helpful to introduce an interim procedure before practicing the procedure described above. The interim procedure involves creating a memory aid: representing the second addend with a finger pattern on the writing hand. The child then uses the nonwriting hand to count on the second addend as described above. When the finger patterns on each hand match, the child stops counting.

7. *COL remedial efforts should focus first on prerequisite skills.* Some children persist in counting all, either concretely or mentally. COL is an extension of the *N*-after rule for one problems. Ensure that children can automatically add one before proceeding with efforts to cultivate COL with non-one problems. Unlike one problems, non-one problems require a keeping-track process. If children are mentally counting all (using CAF or CAL), they already have this prerequisite. If children are still using concrete procedures, they need to learn how to keep track.

8. *COL remedial efforts should focus on helping the child to see unnecessary effort.* Probably most children do not spontaneously count on because they do not recognize that counting up to the first term produces the same result as simply stating the cardinal designation of the first term

FIGURE 8.4: Using Chisenbop Finger Patterns to Keep Track

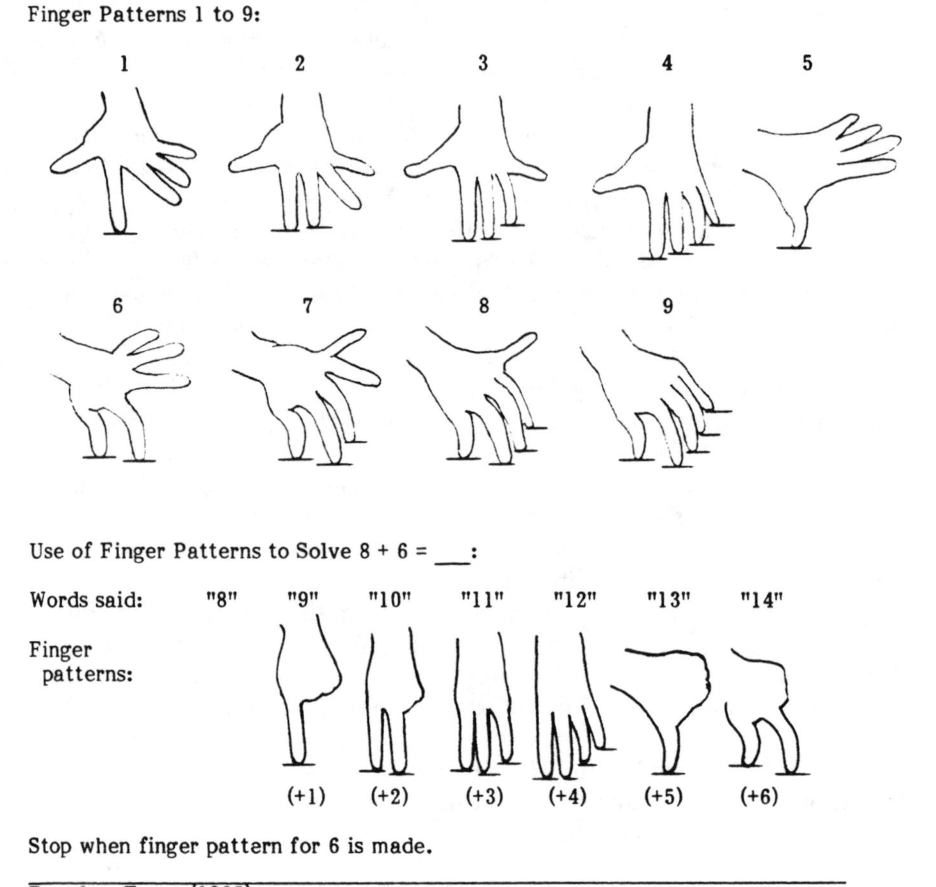

Based on Fuson (1985).

(Baroody & Ginsburg, in press). Figure 8.5 illustrates a teaching method devised by Secada, Fuson, & Hall (1983) that is often successful in quickly helping children to see that counting out the first addend is redundant to stating its cardinal designation.

Subtraction

1. *Ensure mastery of prerequisite skills for counting down.* If a child is having difficulty computing differences with subtrahends of two or more,

FIGURE 8.5: Teaching Counting On

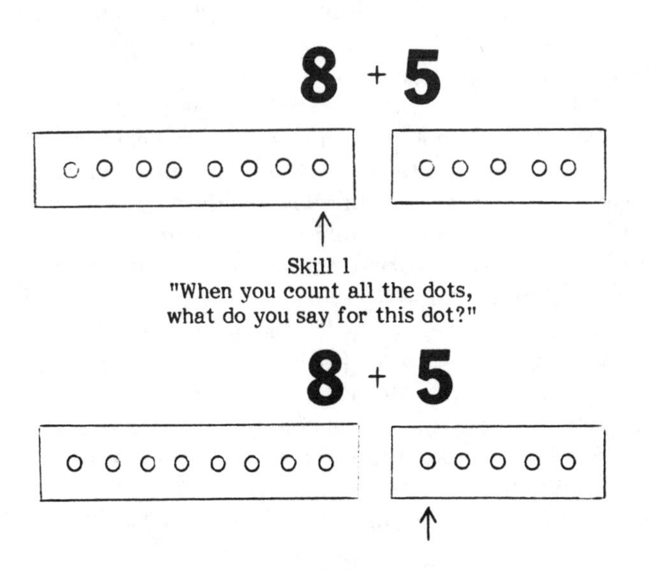

Skill 2
"When you count all the dots,
what do you say for this dot?"

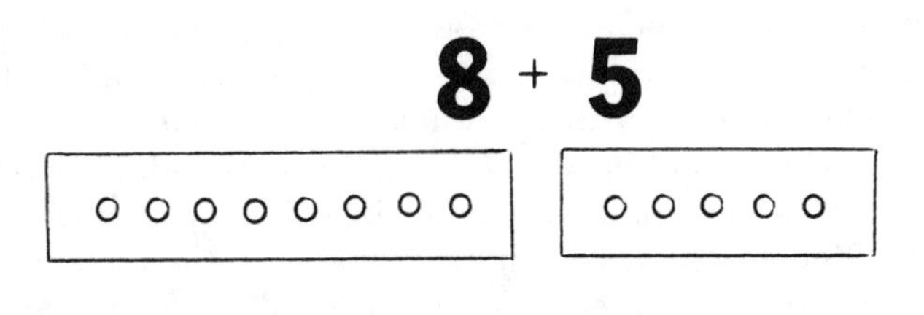

Count On
"How many dots are there?"

Based on Fuson (1985).

the child may be having a problem mastering counting down. In such cases, it is important to check the prerequisite skills for counting down. First, if children lack mental facility with $N-1$ differences (the first step in a counting-down procedure for non-one problems), they will not be able to mentally subtract with minuends of two or more. Therefore, ensure that children can efficiently compute $N-1$ differences first.

Second, if children do not know how to count backward, they will not be able to extend their mental $N-1$ procedure into a genuine counting-down procedure. Moreover, to count down, children must not only be able to count backward but also be able to do it with ease. Otherwise, the demands of simultaneously executing two processes in opposite directions may be overwhelming (Baroody, 1983a). If the backward count is not at least fairly automatic, attention will be split between this count and the simultaneous process of keeping track of the subtrahend. This divided attention may result in an error in the backward count, the keeping-track process, or both. For example, when counting backward is an effort, children sometimes leave out a term in the backward count (e.g., 19 – 5: "19, 18 [–1], 17 [–2], 16 [–3], *14* [–4], 13 [–5]—13"). One second grader experiencing difficulties in mathematics started to solve 19 – 5 by counting backward but had to pause to think what came before 16. As a result, he lost track of the forward count, "19, 18 [one down], 17 [two down], 16 [three down], ah, ah, ah, 15 [ah, *three* down], 14 [four down], 13 [five down]—13." Because counting backward was not effortless, one learning-disabled fifth-grade boy, Adam, lost hold of both counts when attempting to compute 19 – 5: "19, 18, 17, 16, ah, ah, 17, 18, 19, 20, 21."

If a child is unable to count backward or do so with facility, remediation should focus on this informal counting skill. Because counting backward from 20 is often more difficult for primary school children than counting backward from 10, children who rely on a counting-down strategy may not experience learning difficulties immediately but may fall behind when their subtraction assignment includes minuends in the teens (or beyond). In such cases, remediation should focus specifically on counting backward from 20 to 10.

Until counting backward becomes automatic, children might be encouraged to practice their informal counting-down procedure with a number list. Some children discover that the classroom's clock is a useful number line for computing. By using a number list, the child is relieved from the burden of generating the difficult backward count and thus can focus attention on the keeping-track process (see Figure 8.6). When they are mastering counting down, children should *not* be discouraged from using concrete models. Rather, children should be encouraged to use their separating-from strategy. They can also be encouraged to continue mastering the $N-1$ combinations.

FIGURE 8.6: Counting Down with a Number List or Clock to Solve 9 - 3 = ___

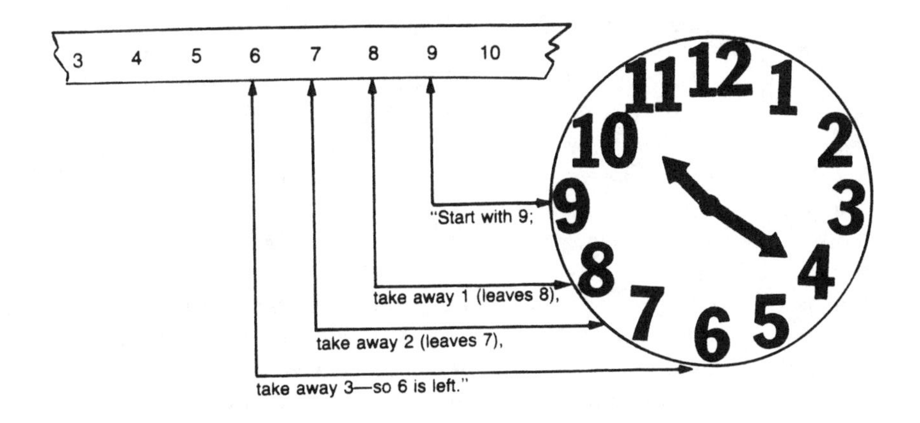

2. *Foster efficient keeping-track processes*. Though an inability to count backward efficiently can undermine the simultaneous processes required by counting down, other factors may also prevent or impair this cognitively demanding procedure. To count down, children must be aware of the need to keep track of the number of times they must count backward and have a preplanned means for doing so (Steffe et al., 1983). Because they do not think to keep track, some children do not know where to stop and hence either proceed until they exhaust the backward sequence or tend to respond incorrectly (Fuson et al., 1982). Indeed, Carpenter and Moser (1982) found that only half of the first graders in their sample could count backward a specified number of steps, and, as a result, a counting-down procedure was used infrequently. Remedial efforts can begin with having children count backward a specified number of times. Begin with counting backward one or two times and work up to more difficult efforts. Then explicitly point out the need for keeping track when computing and how to accomplish this process. (This can be done with concrete models as illustrated in Figure 8.6).

Even after learning a keeping-track process some children feel compelled to rush through the counting-down procedure (often to avoid the stigma of counting). This haste may well result in losing track of one or both of the simultaneous counts or processes. In such cases, the child needs to be reassured that counting down is a clever and common strategy and that accuracy is as important as speed. Another frequent error involves starting the keeping-track count too soon—with the cardinal designation of the minuend (e.g., 17 − 3: "17 [−1], 16 [−2], 15 [−3]— 15"). This error will produce a pattern that is easily detected: The child's answer is nearly always one more than it should be. In such instances, the counting-down strategy can be modeled for the child. It should be emphasized that the "counting process" does not begin with the larger number (minuend).

3. *Encourage the development of counting up and flexibly choosing the more efficient computing procedure.* If children rely exclusively on counting down, they may be accurate with small but not larger problems. As assignments involve larger values, the backward counts become longer and more prone to error. Thus it might be helpful to encourage the child to learn a counting-up procedure and use it when it would be easier to use this procedure instead of counting down. One method for introducing a counting-up procedure is described in Example 8.1. Children who have already discovered counting up might be encouraged to discuss their procedure and when they find it most useful. Some children may benefit from explicit instruction concerning the interchangeability of counting down and counting up. However, note that some children may not pick up on this procedure because it does not correspond to their informal notion of "take away." In such cases, do not pursue the matter. It may only confuse the child, who may eventually discover the strategy later anyway.

Multiplication

1. *Make explicit the connection between multiplication and repeated addition.* Difficulties with basic multiplication can occur because children do not see the connection between this new operation and their existing knowledge. Sometimes remediation simply entails helping the child to make this connection. Consider the case of Ken, a third grader placed in his school math lab because of learning difficulties (Baroody, 1986). His math-lab teacher asked me to see Ken and explained that he had no concept of multiplication. Given several simple multiplication problems such as 6 × 2 = __ and 3 × 3 = __, Ken did not seem to have any idea of what to do. He indicated that he was convinced multiplication was too difficult for him. A counting procedure for multiplication was demonstrated to Ken: 4 × 3 is four fingers counted (without restarting) three times (1, 2, 3, 4; 5, 6, 7, 8; 9, 10, 11, *12*). Ken exclaimed: "Oh, so that's all that multiplication is!" When introduced in an informal way, multiplication made sense, and Ken quickly learned to compute products. Indeed, he quickly began to find shortcuts for the taught procedure. Later, for instance, he responded to 7 × 2 immediately with 14. Upon questioning, he explained that he had used the known combination 7 + 7 = 14.

Sometimes multiplication difficulties are more deeply rooted, and a more concrete approach is required (Baroody, 1986). For example, Adam, a genuinely learning-disabled child, presented a more serious challenge than Ken. Adam was shown the same informal mental procedure that was used successfully with Ken. However, the mental procedure, if anything, was confusing to Adam. Thus a concrete strategy that more clearly modeled the repeated addition of the units was tried. The problem $4 \times 3 = \_$, for example, was solved by putting out three groups of four blocks and then counting the blocks.

2. *Explicitly encourage skip counting, especially for large, difficult-to-compute combinations.* Children usually multiply small numbers (up to 5×5) with little difficulty but often experience difficulty with problems that involve the numbers 6 to 9. For small-number problems they can make a finger pattern of the multiplicand on one hand and use the fingers of the other hand to keep track of the number of times the multiplicand has been counted (e.g., for 4×3, put up four fingers on the left hand and keep track of three counts with fingers on the right hand). For larger problems, such as 6×5 or 5×6, there are an insufficient number of fingers for this common computing procedure.

To get around this difficulty, Wynroth (1969/1980) suggests a vertical keeping-track method. For 7×6, if a child chooses to tally sevens, seven fingers are put up and counted. A 7 is then recorded on lined paper. The seven fingers are counted again (beginning with "eight"), and 14 is then recorded on the lined paper. The process is continued until the child has made six entries as shown below.

7
14
21
28
35
42

For some children it may help to label the multiplicand and the multiplier. Wynroth (1969/1980) suggests that, to start with, the child should record and circle the multiplicand (the number to be counted). In the example above, a circled 7 would have been placed at the top of the column before the child began counting. Moreover, it may help to have a first column that specifies the multiplier. In the example above, a column of numbers 1 to 6 could be placed to the left of the existing column. The uncluttered procedure described and shown in the previous paragraph may actually be less confusing for some children—especially special education children.

(continued on p. 154)

EXAMPLE 8.1

Teaching Counting Up: The Balance Activity

Level 1: Using objects as a concrete support.

Step 1: As diagrammed below, set up the balance with an unequal number of weights (of different color) on each side.

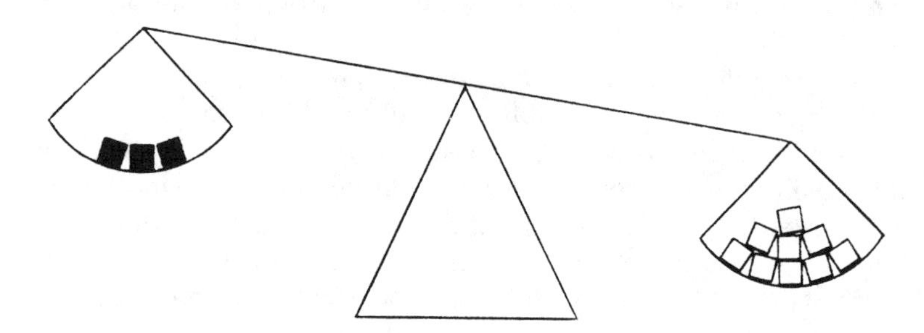

Step 2: Point out the number of weights on each side ("3" and "9") and ask the child to determine how many weights have to be added to the light side (the side with three) to make the balance even.

Step 3: If the child needs help, suggest adding blocks. (If available, use a third color for contrast.) Add blocks to the light side until the balance is even.

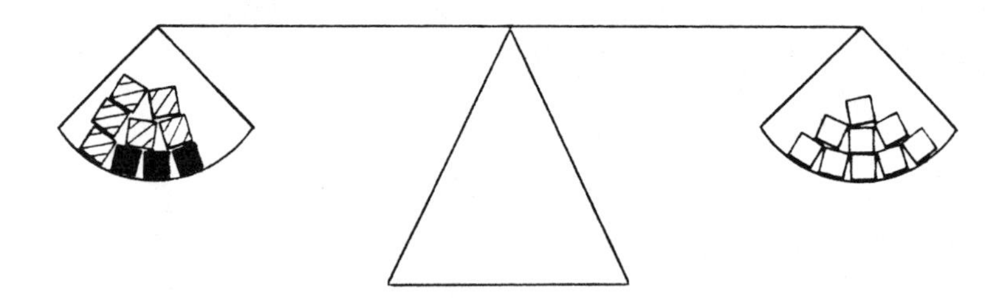

Step 4: Ask the child to count how many blocks were added to the light side to make it even. Later, when the child is ready, Steps 3 and 4 can be combined. The child can simply count the blocks as they are added.

EXAMPLE 8.1 (continued)

Level 2: Using a number list as a semiconcrete support. This can usually be introduced after the child has mastered the counting-up procedure using concrete objects. Initially, the child may have a third weight at various parts along the light side. Eventually the child should be encouraged to just count up from 3 to 9 on the number list.

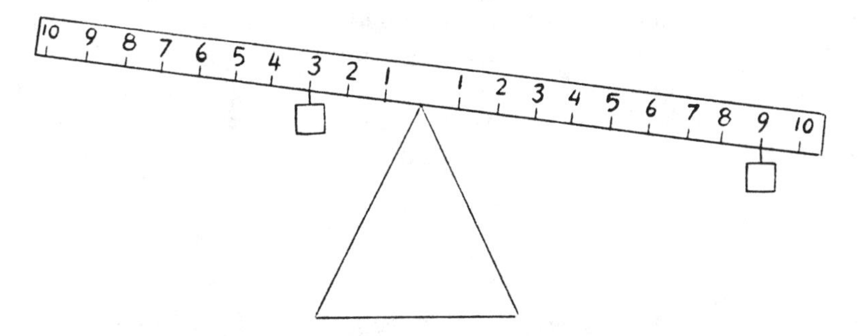

Level 3: Using weighted numerals to encourage mental counting up. After the child has computed an answer, require the responder to hang a numeral corresponding to the answer on the light side as a check.

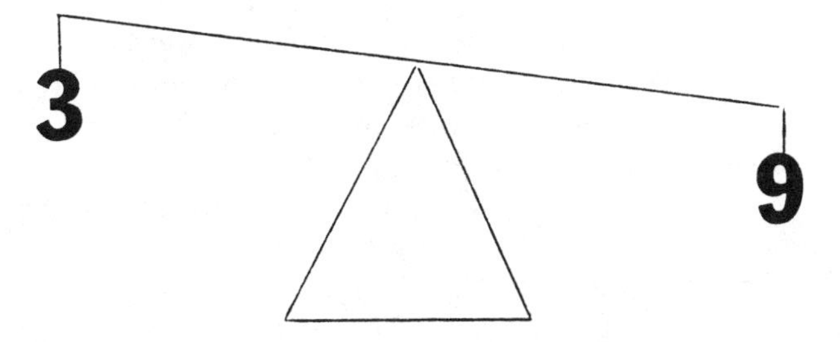

Optional enrichment activity: Note that, for all three levels, children can practice making predictions or estimates, which they can then check by computing. They can determine the accuracy of their estimate by computing the difference between the estimate and their computed answer. By recording the accuracy of their estimates for a problem, they can note changes in their prediction accuracy over time.

This vertical keeping-track method has a number of virtues. If children lose track while calculating, they can simply count up the number of entries they have made (and continue to count on from there). Moreover, children can reuse the record. For problems with a smaller multiplier, such as $7 \times 4 = \_$, a child simply has to count down the seven column to the fourth entry, and the answer is already recorded. Note that this vertical keeping-track method is consistent with children's natural tendency to use skip counting. For problems with a larger multiplier, such as $7 \times 7 = \_$, a child can go down the seven column to the last (sixth) entry and count on seven from there (and make a new entry). Eventually a child will have a complete record of all the basic multiplication combinations.

SUMMARY

Before mastering the basic number combinations, children rely on counting-based computing procedures, which initially require concrete objects such as fingers and blocks. Typically, children naturally gravitate toward using verbal or mental counting procedures to compute. Children readily learn to use their knowledge of the number sequence to efficiently answer $N + 1 = \_$ and $N - 1 = \_$ questions. Children's informal understanding of arithmetic guides their construction or invention of concrete and mental computing procedures. Because children view addition as adding *more* to something, reversed problems such as $5 + 1$ and $1 + 5$ and $3 + 5$ and $5 + 3$ are seen as different problems. As a result, children may feel compelled to compute the sum of $1 + 5$ even though they know that $5 + 1 = 6$. Children soon discover that $1 + N$ and $N + 1$ produce the same sum and that the efficient number-after rule applies to $1 + N$ as well as $N + 1$. Eventually children learn that addend order does not affect the outcome of non-one problems either (e.g., $3 + 5 = 5 + 3$). Mental computing is cognitively demanding because children must keep track of how far to count while they are counting. Thus increasingly larger problems entail an increasingly more complicated keeping-track process, and there is a real incentive for children to invent new computing procedures to minimize this mental labor. Thus, nonconceptual factors as well as conceptual growth play a role in the development of informal computing procedures. Informal computing difficulties can occur because component counting skills or keeping-track processes are not adequate or efficient.

Part III

FORMAL MATHEMATICS

CHAPTER 9

Arithmetic Concepts

How do children learn formal mathematics in a meaningful way? How can instruction foster the meaningful learning of mathematics? In what ways does informal knowledge limit children's learning of formal mathematics? How can a teacher best deal with psychological blind spots?

THE NEED FOR MEANINGFUL LEARNING

Adam's Mechanical Checking Procedure

In his class for learning-disabled children, Adam had been taught the standard procedure for checking subtraction: Add the difference to the subtrahend and compare this sum with the minuend. If the sum and the minuend are the same (e.g., $35 - 17 = 18$; $17 + 18 = 35$) then the subtraction should be correct. Adam perfunctorily completed a whole page of problems, first subtracting and then adding the result to the bottom number of the subtraction problem. Though Adam had learned the mechanics of checking, he did not really understand why the procedure worked. To Adam, checking subtraction by adding was a senseless procedure that simply added work. Because it was a rote and pointless procedure, Adam did not spontaneously use the checking procedure unless explicitly instructed to do so. Even then, he did not use it effectively to accomplish its intended aim.

The rationale for the procedure had not been made clear to Adam. Indeed, even if it had been given, Adam probably would not have profited from an explanation of the procedure's rationale: Addition and subtraction are related operations in that one can undo the other. Thus the effects of subtracting 17 from 35 can be undone by adding 17 to the outcome.

Using the magic task described in Chapter 7, a diagnostic interview found that Adam did not have the informal knowledge to grasp the rationale of the task. When one block was secretly added to (or removed from) a set, Adam had no difficulty indicating that a block had been added (taken away) and that the original set could be reformed by taking away the extra block (adding one block). However, Adam could not extend the principle to larger changes. For example, he was shown a pan with seven blocks. Then the interviewer ("Mr. Magic") covered the pan with a cardboard sheet and surreptitiously removed two blocks.

ADAM: Five.
INTERVIEWER: I thought there were seven in there. What did Mr. Magic do?
ADAM: He took two.
INTERVIEWER: I took away two. And how would I restore what was there originally? How would I get back to seven?
ADAM: In a magical way.
INTERVIEWER: How can I get from five back to seven again?
ADAM: Six, seven, [fingers move two times]: two.

Like preschool children (e.g., Gelman, 1977), Adam had a general sense that addition could be undone by subtraction. However, beyond additions (reductions) of one, he did not have a precise idea of how addition and subtraction canceled each other. That is, he did not seem to realize that the subtraction of two could be undone by the addition of two. Because he did not have an exact understanding of this relationship beyond ± 1, he had to compute to determine how many had to be taken away (added) to restore an original set when the change involved two or more.

Like Adam, is it common for schoolchildren to learn formal procedures that they do not understand? What are the consequences of such learning? How could Adam have learned the conceptual basis for the checking procedure? How could Adam have been helped to see the connection between this concept and the skill of checking?

Knowledge of Form and Understanding

It is useful to distinguish between two types of mathematical knowledge: knowledge of form and understanding (e.g., Hiebert, 1984). Knowledge of form involves learning the names and shapes of symbols (e.g., how to read and write the numerals), the procedural rules for using symbols (e.g., the larger number is placed at the wide end of the greater-than symbol, as in $5 > 3$), and the procedural rules for manipulating symbols to solve problems (e.g., the carrying algorithm for multidigit addition). Understanding involves constructing concepts or finding connections. For example, from informal experience, children construct a concept of addition as an incrementing process. In school, children assign meaning to the plus sign (+) by connecting the symbol to their concept.

INFORMAL KNOWLEDGE AS A BUILDING BLOCK

Concept Learning

Children's informal mathematics can be an important source of mathematical understanding. The application of informal skills can and often does lead to important insights. A prime example is the discovery that some problems that do not look alike can actually have the same sum.

THE DISCOVERY OF THE COMMUTATIVITY PRINCIPLE. Children do not naturally assume that commuted problems, such as $2 + 4 = \_$ and $4 + 2 = \_$, or other nonidentical but equivalent problems, such as $4 + 4 = \_$, $5 + 3 = \_$, and $6 + 2 = \_$, have the same sum. As we saw in Chapters 7 and 8, children tend to view addition as an incrementing process. As a result, they interpret $4 + 2$ as "four and two *more*" and $2 + 4$ as "two and four *more*." Moreover, because they view $2 + 4$ and $4 + 2$ as different problems and because they cannot foresee that the outcomes will be the same, children just naturally assume such problems have different sums (Baroody & Gannon, 1984). Children are often surprised to discover that different-looking problems—both the commuted variety (e.g., $4 + 2$ and $2 + 4$) and the noncommuted variety (e.g., $4 + 4$, $5 + 3$, $6 + 2$)—can actually produce the same sum. In the protocol excerpts below, note how a kindergartner's confidence in her knowledge of commutativity evolved as a result of informal experience.

INTERVIEWER: This is 6 and 1 [pointing to $6 + 1$ typed on a $3'' \times 5''$ card] and this is 6 and 1 [pointing to a second card 2.5 cm below the first]. Do you think they would add up to the same or different answers?
KATE: [Immediately] Same.
[To $2 + 5$ and $2 + 10$, immediately] Different.
[To $2 + 4$ and $4 + 2$: No response.]
INTERVIEWER: [Turns cards over.] Just tell me quickly what you think.
KATE: [Pause] Same.
[To $0 + 2$ and $10 + 10$, immediately] Different.
[To $3 + 1$ and $0 + 1$, immediately] Different.
[To $5 + 3$ and $3 + 5$] I can't tell. [Pause.]

INTERVIEWER: Quickly, what do you think?
KATE: Same
[To 2 + 2 and 5 + 2, immediately] Different.
[To 4 + 6 and 6 + 4] That's the one I got so much trouble over. Same?

Initially, unlike identical problems that clearly implied the same sum and problems that differed by one term that clearly implied different sums, commuted problems—with their like but differently ordered terms—were a cause of uncertainty. With the next three commuted pairs, she paused only once before responding. In a second session, she responded quickly to all six commuted pairs with: "The same." Asked why she thought two commuted items (3 + 4 and 4 + 3) would be equivalent, Kate responded, "Because same numbers in different places look like they add up to the same." Kate, then, seemed to have cautiously concluded that addend order did not matter. In a third session, Kate was asked to compute the sum of 6 + 4. After recording her answer, the interviewer wrote down 4 + 6 directly beneath the previous problem and asked Kate if 4 + 6 would add up to the same thing as 6 + 4 or something different. Kate immediately indicated: "Yes." Asked why, she noted: "I figured it out when I counted when we played the other game." Thus it appeared that Kate had—from her computing experience—constructed a principle that was applied consistently, generally, and with confidence.

DISCOVERY BY CHILDREN WITH LEARNING DIFFICULTIES. Even children with learning difficulties can discover—and enjoy discovering—mathematical relationships. Despite his severe learning disability, Adam rather quickly picked out the regularity of commutativity. In the context of a baseball game, Adam was presented with a series of addition problems. Sometimes a problem (e.g., 5 + 8 = __) was followed by its commuted counterpart (8 + 5 = __) and sometimes by a different problem (5 + 0 = __). After Adam computed the sum for a problem, his answer was recorded (e.g., 5 + 8 = 13), and placed in a discard pile—face up—just off to his right. If the next problem was a commuted counterpart, Adam could avoid recomputing the sum by turning his eyes or head to view the equation in the discard pile. Because the number of bases his hitter could take was dependent on how quickly he responded (1 to 3 seconds was a home run, 4 to 6 seconds was a triple, 7 to 9 seconds was a double, and 10 seconds or more was a single), there was an added incentive to avoid laborious calculation.

Initially, Adam tediously calculated every problem—including commuted counterparts. Apparently he did not realize the shortcut that commutativity afforded. The second time he played the game, Adam missed two more opportunities to exploit this relationship (with the pairs $4 + 2$ and $2 + 4$ and $6 + 4$ and $4 + 6$). Then, after computing the sum of $4 + 3$, he was given $3 + 4$. He looked at the previous equation and quickly responded, "Seven!" Thereafter he regularly used the commutativity principle to minimize effort—even with larger pairs, such as $14 + 7$ and $7 + 14$ and $15 + 13$ and $13 + 15$. It appears that such discovery-learning exercises can also sensitize mentally handicapped children to commutativity and prompt them to exploit this relationship.

DISCOVERING THE RELATIONSHIPS BETWEEN ADDITION AND SUBTRACTION. Though preschoolers typically understand that the subtraction of one is undone by the addition of one (or vice versa), children may not discover a general *inverse principle* until after having had some computing experience with non-one addition and subtraction problems. For example, by taking away two of three fingers, a child might imagine that replacing the two fingers would recreate the original set of three (i.e., $3 - 2 + 2 = 3$). Through such computing experiences and reflection, a child might realize that the subtraction of any given N can be undone by the addition of N (or vice versa).

Moreover, such activities may lead to the discovery of a related but less apparent connection between addition and subtraction. Addition and subtraction are related by the *complement principle*: Subtracting a part (an addend) from a whole (the sum) yields the other part (addend). For example, a child might find that two fingers plus another makes three and that subtracting one part from this sum yields the other part (i.e., $2 + 1 = 3$; therefore, $3 - 2 = 1$ or $3 - 1 = 2$).

Connections

Informal mathematics is a crucial basis for assimilating formal mathematics (Ginsburg, 1982). As we saw in Chapter 8, children can assign meaning to the formal symbols and procedures for an operation by connecting these forms to their informal knowledge of an operation. The same principle holds true for much of primary-level formal mathematics (Baroody, 1986).

Teachers and textbooks define important mathematical relationships in words and symbols. Specifically, commutativity is depicted as $a + b = b + a$ or less abstractly in instances such as $5 + 3 = 3 + 5$. The inverse principle is exemplified by expressions such as $5 - 3 + 3 = 5$ or $5 + 3 - 3 = 5$. The equivalence of repeated addition and multiplication is illustrated by instances such as $4 \times 3 = 4 + 4 + 4$. Children often develop an informal understanding of such key mathematical relationships before exposure to the form in school. For example, research (Baroody et al., 1983) suggests that most first graders discover commutativity before this principle is formally introduced. In fact, if given sufficient computational experience, even many disadvantaged first-grade children discover the principle without benefit of formal instruction (Baroody et al., 1982). Research suggests that an inverse principle is also informally understood by first graders (Bisanz et al., 1984; Gelman, 1977). Moreover, first graders can readily see the kinship between multiplication and addition (Wynroth, 1969/1980). When formal definitions and representations are introduced in terms of children's informal knowledge, these forms are more meaningful and more readily mastered.

Difficulties

Instruction based on absorption theory focuses on form: written symbols and procedures. Moreover, it introduces formal mathematics at a very quick pace and without regard for children's informal knowledge. Though knowledge of form can be rotely memorized relatively quickly, the development of understanding is an involved and often lengthy process. Moreover, as children proceed through school, mathematics can become increasingly unrelated to their experience. As a result, children often acquire knowledge of form without understanding. That is, children often learn facts, definitions, and procedures without really understanding this formal mathematics. Unfortunately, rotely learned knowledge tends to be applied mechanically. Furthermore, a rote approach may result in an incomplete learning or retention of formal mathematics. Efforts to rotely memorize formal mathematics may even fail completely.

INSUFFICIENT INFORMAL EXPERIENCE AND EXPLORATION. Because of the quick pace of mathematics instruction and because informal mathematics is sometimes overlooked or discouraged, children may not have the opportunity to discover mathematical relationships in a meaningful manner. Children with poorly developed informal arithmetic skills and little informal calculating experience are far less likely to discover basic relationships such as commutativity—not to mention less obvious relationships such as the addition-subtraction complement principle (Baroody et al., 1982). Adam, for example, apparently had not discovered a general inverse principle or the addition-subtraction complement principle. These concepts underlie an add-to-check-subtraction procedure. Because he did not have the basis for understanding its rationale, Adam only mechanically learned and used the checking procedure.

GAPS BETWEEN FORMAL AND INFORMAL MATHEMATICS. Many children have difficulty understanding and learning formal mathematics because they fail to connect the formal symbolism to their practical, everyday mathematics. Even demonstrations with concrete objects, illustrations with pictures, or examples with the symbols may not promote assimilation unless children see the connection with their informal mathematical knowledge.

A gap betwen informal and formal mathematics can result in rote memorization of school mathematics. In theory, children are introduced to mathematical expressions, such as $4 \times 3 = 4 + 4 + 4$, to help them learn key mathematical relationships. Unfortunately, when such expressions are not related to their informal knowledge, the formal expression may simply be rotely memorized (if learned at all). Thus instead of helping a child to understand an important mathematical idea, instruction that fails to make connnections with informal knowledge may succeed only in imposing incomprehensible and useless facts upon children.

A gap between formal and informal knowledge frequently leads to difficulties with written arithmetic. Despite their informal knowledge of arithmetic, some children have difficulty when written, symbolic arithmetic is introduced. Even though children can solve verbal arithmetic problems, such as, "How much is five take away three?" they may answer the very same problems incorrectly on their written assignments. If children do not relate the minus symbol to their informal understanding of subtraction (as "take away"), they will fail to employ their informal subtraction procedure. Unsure of the sign's meaning, some children simply add.

EDUCATIONAL IMPLICATIONS: DESIGNING MEANINGFUL INSTRUCTION

A central objective of initial mathematics instruction should be to cultivate understanding: promote concept learning and link knowledge of form with concepts. Meaningful instruction of mathematics takes into account and builds upon children's informal mathematics. This entails helping children to see how formal symbols and procedures are connected to and expand upon their existing practical knowledge of mathematics. Some specific recommendations follow.

1. *Develop a solid base (informal understanding) before introducing written symbolism*. Children need an extended period of time with concrete objects and concrete problems to develop an understanding of number, the arithmetic operations, mathematical principles, and place value before they are confronted with written assignments. Many children have

some concrete understanding of number, addition, and subtraction when schooling begins. These children should be encouraged to pursue their informal mathematics so as to discover important mathematical relationships, even as written assignments are introduced. For children with little or no informal understanding of number and arithmetic—especially environmentally deprived and special education children—great care and time may be needed to strengthen these fundamental concepts. For more advanced concepts such as multiplication, the commutativity principle, and positional notation, practically all children can profit from a period of informal exploration and practice. In this way, formal mathematics can be connected to something meaningful. For example, instruction on how to check subtraction should have been introduced after Adam had the opportunity to discover general inverse and addition-subtraction complement principles. This would have increased the chances of learning the skill in a meaningful manner.

2. *Structure informal calculational experiences to promote discovery learning.* By carefully structuring exercises, children—even children with learning difficulties—can be led to discover important mathematical relationships through their informal computing efforts. Efforts to help children "see" principles should start with small, easily calculated combinations. Such calculations entail less attention and increase the child's chances of noticing a regularity. For example, to encourage the discovery of the addition-subtraction complement principle, Adam could have been given sequences of problems with small numbers (e.g., $2 + 1 = 3$, $3 - 1 = 2$, $3 - 2 = 1$) or familiar combinations (e.g., $5 + 5 = 10$, $10 - 5 = 5$).

Consider the case of Ken, a third grader in a remedial math class (Baroody, 1986). Though he had no difficulty with 6×3 (six counted three times), Ken seemed overwhelmed by the problem 3×6. He indicated that he had tried to count three six times but was unable to do so. Ken needed to see that 3×6 was equivalent to 6×3—that multiplication is commutative. Instead of simply telling Ken the commutative law, which he might or might not have understood, I gave him an opportunity to discover the principle informally. To maximize the chance of his noticing that the order of the terms does not affect the outcome, I started with easily calculated problems. After quickly solving 3×2 (three counted twice) and 2×3 (two counted three times), Ken concluded: "Oh, they're the same." He then used the abstracted principle of commutativity to solve more difficult problems such as 3×6 (now solvable by using the procedure for 6×3) and 2×100 (the equivalent of 100×2, which for Ken was the equivalent of $100 + 100$).

3. *Help children to see that formal symbolism is an explicit expression of their informal knowledge.* To help children see the connection between formal mathematics and their informal knowledge, point out how mathematical symbols capture and summarize what they already know. Help the

child to see that formal symbols and expressions are just a means of making clear what we believe about mathematics. The following case illustrates such an effort.

Informally or implicitly, Adam treated repeated addition and multiplication as equivalent. He used skip counting to solve both $5+5+5+5+5$ and 5×5. Adam could also use the standard algorithm to calculate the products of written problems like

$$\begin{array}{r} 432 \\ \times \quad 2 \\ \hline \end{array} \qquad \begin{array}{r} 123 \\ \times \quad 5 \\ \hline \end{array} \qquad \begin{array}{r} 121{,}152 \\ \times \qquad 2 \\ \hline \end{array}$$

Nevertheless, formally or explicitly, he did not realize that $432 + 432 = 432 \times 2$ or that $5+5+5+5+5 = 5 \times 5$.

> INTERVIEWER: [Presents five piles of five coins.] I bought five packages of coins, each package has five coins in it. How many coins do I have?
>
> ADAM: [Without hesitation.] Twenty-five.

Asked to represent the problem in written form, Adam wrote $5+5+5+5+5$. Asked to show other ways to write the problem, he indicated $10 + 15$. Apparently, he did not view $5 \times 5 = 25$ as applicable.

After it was pointed out that $5+5+5+5+5$ could also be written as 5×5 and that $4+4+4+4+4$ could also be written as 4×5, Adam continued to have some difficulty. He was given three piles of five coins each. Adam immediately indicated that the total would be 15. Asked to represent the problem, he wrote:

$$\begin{array}{r} 5 \\ 5 \\ +5 \\ \hline \end{array}$$

Asked to say $5+5+5$ in another way, Adam responded, "Five times one? [Pause.] Five times five?" After it was noted that there was one pile of five, another pile of five, and yet another pile of five, Adam responded, "Five times three." Next Adam was given:

$$\begin{array}{r} 5 \\ 5 \\ 5 \\ 5 \\ 5 \\ 5 \\ 5 \\ +5 \\ \hline \end{array}$$

He counted by fives and correctly remarked, "Forty." Asked another way to write the problem, Adam wrote 5×4. Asked how many fives there were, Adam counted the eight fives and then wrote: 5×8.

By explicitly linking the written form for multiplication to repeated addition and his informal calculating procedure (skip counting), Adam slowly began to comprehend formal expressions such as $5 + 5 + 5 + 5 + 5 + 5 + 5 + 5 + 5 + 5 = 5 \times 10$. Presented an addition problem with 10 fives, Adam commented, "That's easy," and proceeded to count by fives to 50. "So," he spontaneously concluded, "that's going to be 5×10." Asked to solve an addition problem consisting of three sixes, Adam calculated a sum of 18 and then noted that the problem could also be written as 6×3.

To reinforce the explicit connection between repeated addition of like terms and multiplication, Adam was required to compute scores in two ways. Adam greatly enjoyed target practice, and scoring targets served as a naturally entertaining device to tally repeated instances of a number. For example, if Adam hit the four ring of the target three times, he was asked to represent the score both as multiplication ($4 \times 3 = 12$) and as repeated addition ($4 + 4 + 4 = 12$).

4. *Sequence formal instruction to exploit children's informal knowledge.* Curriculum organization should take into account children's informal mathematics. Curriculum designers and textbook publishers often focus on external factors to sequence instruction: custom, the structure of formal mathematics, and task analyses (a logical analysis of a topic's prerequisite and component skills). However, no matter how carefully these external factors are considered, instruction will not be efficient if it does not take into account internal factors. In the elementary years, this means considering children's informal knowledge.

For example, commercial textbook series typically sequence multiplication instruction by problem size. The *Holt School Mathematics* (Nichols et al., 1978) text introduces one through five combinations in Chapter 7. Two and three combinations are practiced first, one and zero combinations second, and four and five combinations last. Ten combinations are not introduced until Chapter 9, after the six through nine combinations are introduced in Chapter 8. Logically, it makes sense to proceed from small to large.

Yet introducing the five combinations after threes and fours and introducing the ten combinations after all single-digit combinations have been taught overlooks an important psychological fact. As with addition, children typically use informal procedures based on counting to calculate products (e.g., 5×3: "5, 10, 15"). Because children learn to count by twos, fives, and tens before they learn to count by threes, fours, and so forth, the two, five, and even ten combinations are relatively easy for most children to work with. Thus practice with both five and ten combinations need not be postponed. In fact, the five-times family is psychologically an ideal candidate for introducing multiplication (Thorndike, 1922). It presents the idea that multiplication is repeated addition of a like term more effectively than the one-times or even the two-times combinations.

A Case in Point: The Wynroth Curriculum

Meaningful learning of formal mathematics requires a readiness to learn, instruction that can be assimilated, and time for the assimilation to occur. The Wynroth (1969/1980) curriculum, designed for children in grades K–4 or those having learning difficulties with basic mathematics, is an interesting example of a curriculum that takes into account these key psychological considerations.

A SOLID, INFORMAL BASIS FOR LEARNING. The Wynroth (1969/1980) curriculum consists of a series of games and worksheets, which children proceed through at their own pace. The games provide the informal basis for understanding and learning formal mathematics. The worksheets serve to teach knowledge of form (symbols and procedural rules) and to extend understanding. The first phase of the curriculum focuses on basic number concepts and skills. The second phase consists of three subsequences that are taught simultaneously: arithmetic operation, missing number, and base ten.

The curriculum ensures readiness for learning formal mathematics in a number of ways. The first phase ("the preworksheet phase") consists entirely of concrete activities: games. Games serve as the vehicles for teaching concepts informally and practicing informal skills. All first-phase games (basic number concepts and counting skills) are mastered before the child begins any worksheets (formal mathematics). This ensures that the child has sound informal number concepts and skills before continuing with more advanced work. Moreover, during the second phase, games are introduced well in advance of written work so that the child has a solid (informal) understanding before dealing with formal symbols. Finally, in both phases, the child proceeds to the next game or worksheet only after mastering previous games or worksheets.

The curriculum cultivates and builds on children's natural or informal mathematics. For example, multiplication is defined as repeated addition: "In order to 'multiply' two numbers, one *adds one of the numbers* to itself as many times as indicated by the *other number*. . . . To teach [four times three] or 'four, three times' say: '*Hold up one number* on your fingers and then *count* those fingers the other *number* of *times*" (Wynroth, 1969/1980,

p. 28). Multiplication is introduced in this manner for two reasons. First, repeated addition is the way the operation is generally used in real-life situations. Second, it is easily understandable because it builds on a child's familiar knowledge of addition.

STRUCTURED DISCOVERY LEARNING. Written exercises, like the one below, are designed to encourage children to look for structure and exploit relationships to shortcut efforts. For example, to solve the last problem in the exercise, a child should recognize that it is unnecessary to calculate each partial sum individually ($3+3$ is 4, 5, 6; $6+3$ is 7, 8, 9; $9+3$ is 10, 11, 12; $12+3$ is 13, 14, 15; $15+3$ is 16, 17, *18*). Six threes is just three more than five threes. Hence 15 (the sum of the previous problem) plus a sixth 3 is 16, 17, *18*.

$$\begin{array}{r} 3 \\ +3 \\ \hline \end{array} \qquad \begin{array}{r} 3 \\ 3 \\ +3 \\ \hline \end{array} \qquad \begin{array}{r} 3 \\ 3 \\ 3 \\ +3 \\ \hline \end{array} \qquad \begin{array}{r} 3 \\ 3 \\ 3 \\ 3 \\ +3 \\ \hline \end{array} \qquad \begin{array}{r} 3 \\ 3 \\ 3 \\ 3 \\ 3 \\ +3 \\ \hline \end{array}$$

CONNECTING FORM TO UNDERSTANDING. Worksheets on multiplication are introduced after a child has mastered informal multiplication procedures. Written work is linked explicitly to children's existing informal concepts and procedures. Symbols, like the times sign ($\times$), are defined in terms of informal concepts and procedures: 4×3 means four three times so use your repeated counting procedure (e.g., count four fingers three times). Moreover, written exercises, like the one described above, are intended to help children see the connection between repeated addition and multiplication. The process of multiplication is introduced as a shortcut for the familiar but relatively cumbersome process of repeated addition. The formal representation of multiplication (e.g., 3×6) is introduced as a more convenient form for repeated addition (e.g., $3+3+3+3+3+3$). Expressions such as $3+3+3+3+3+3 = 3 \times 6$ are tied to equivalent informal processes—the same informal procedures. Thus the equivalence expression makes sense to children.

SEQUENCING INSTRUCTION TO TAKE INTO ACCOUNT INFORMAL KNOWLEDGE. Because of its close psychological kinship to addition and because it is relatively easy to compute informally, the Wynroth (1969/1980) curriculum introduces multiplication after addition instead of subtraction. From a psychological point of view, this sequence is not imperative. Subtraction has a strong kinship to addition that dates back to preschool experiences. Moreover, most children invent at least a concrete procedure for computing differences. Therefore, as long as children's informal approach is respected, there is no need to delay the introduction of subtraction in school.

On the other hand, Wynroth (1969/1980) is on firm psychological ground when he points out that the introduction of multiplication need not be delayed (as long as an informal approach is taken). Indeed, introducing multiplication may be a way of helping children with learning difficulties to gain confidence and a measure of self-esteem. In second and third grade, some children experience considerable difficulty with larger, more-difficult-to-compute basic subtraction combinations and with the renaming procedures for both operations, but especially subtraction. I observed a handful of children who were experiencing so much difficulty in these areas that they had to be assigned to their school's math lab for remedial training. Instruction and practice in these areas was accompanied by chagrined looks and even outright complaining. Yet these children, using an informal approach, had quickly learned to multiply and could efficiently compute products. When the math-lab instruction focused on multiplication, the children became joyful and eager. Because they could readily understand the operation and compute products, these children greeted multiplication with relief. Moreover, multiplication was an aspect of mathematics instruction that made these children feel that they were like their "normal" classmates.

EDUCATIONAL IMPLICATIONS: DEALING WITH BLIND SPOTS

Because school mathematics is assimilated in terms of existing knowledge, informal knowledge can limit or interfere with children's understanding of formal mathematics. Because existing knowledge shapes and thus can distort a child's interpretation of new information, children may only partially understand or even misunderstand instruction involving formal symbols. Moreover, an incomplete understanding of formal mathematics may be reinforced by how the mathematics is taught. Thus, it is important for educators to be aware of common misunderstandings or blind spots caused by children's informal knowledge. In this way, initial instruction can be adjusted to minimize learning difficulties. Consider the following examples.

Subtraction as "Take Away"

As mentioned in Chapter 8, children tend to interpret written subtraction as "take away." This interpretation is a natural outcome of their informal experiences with concrete objects and counting. Moreover, this interpretation is reinforced in school when subtraction is referred to as "take away." Though subtraction can mean "take away," it also has other meanings that children need to learn. Subtraction can have a comparative meaning: the difference between two sets (e.g., How much bigger is a set of 7 than a set of 3?). Furthermore, subtraction can be related to addition by a missing-addend or additive-subtraction approach. For example, $10 - 7 = \_$ can be thought of as: What must be added to a given part (7) to make a given whole (10)—that is, $7 + \_ = 10$.

It has not yet been established how to best overcome children's limited interpretation of subtraction. Some (e.g., Fuson, 1984, in press; Wynroth, 1969/1980) suggest that teachers never refer to the symbol for subtraction (−) as "take away" nor encourage separating from and counting down as informal subtraction methods. In this view, the symbol for subtraction should only be referred to as "minus." Moreover, subtraction should be introduced in terms of a missing-addend concept and a counting-up procedure for three reasons (Wynroth, 1969/1980). (a) Counting up is easier than counting down with minuends of bigger than 1 or 2, and it builds on a familiar counting-on procedure for addition. (b) It may be less confusing to think of subtraction as another form of addition—a highly familiar operation. (c) A missing-addend approach is more applicable to the various meanings of subtraction than a take-away approach.

Psychologically, it is not clear that children can be prevented from initially viewing subtraction as "take away" or that they can assimilate a missing-addend approach. Even after several years of exposure to a missing-addend approach, many children prefer to use a counting-down procedure (Baroody, 1984c). Children often use informal procedures that make sense in terms of their informal concepts rather than adopt procedures taught in school for which they do not have an adequate conceptual understanding (Ginsburg, 1982). Thus, despite the curriculum used, primary teachers can expect some—if not many—of their pupils to rely on a counting-down procedure for mental subtraction.

It may be more productive to accept children's natural interpretation of subtraction as "take away" and supplement it with other meanings of subtraction. A teacher can introduce children to, say, a difference meaning of subtraction by an activity like that described in Example 9.1. After children have had an opportunity to use their own informal method to

EXAMPLE 9.1

Practice with the Comparative Meaning of Subtraction

DIFFERENCE GAME

Objectives: (1) Difference meaning of subtraction
(2) Practice number-neighbor comparisons
(3) Practice computing difference by counting up
(4) Master basic subtraction combinations

Materials: Dice with 0 to 5 or 1 to 6 dots for the most basic game, dice with numerals 0 to 5 or 1 to 6 for children ready to count up mentally, or spinner or card with 1 to 10 dots or numerals 1 to 10 for more advanced players

Instructions: The game is played in pairs. Each player obtains a number (by rolling a die, spinning a spinner, or drawing a card). The players compare their numbers. The player with the larger number wins. The number of points the winner gets is determined by how much the player beat his opponent. For beginners the difference can be computed by separating from. For example, if Jason's card has five dots and Karen's card has three, Jason would be instructed to cover up the first three dots on his card and count up the remaining (two) dots to determine his score. If Jason has mastered this procedure and is ready to mentally count up, he would be instructed to begin with Karen's number 3 and count up from there to his number 5 to find the difference: "four [that's one more], five [that's two more]—two points."

figure differences, a teacher can point out that the formal symbols used to represent take away can also represent difference problems. For example, the expression $7-5=\_$ can be used to represent the difference between five and seven as well as seven take away five. A teacher can also point out the various meanings of subtraction by using word problems (see Table 9.1). The teacher can help children to see that the formal symbolism can represent and thus connect the different meanings of subtraction.

Equals as "Produces"

Children are particularly prone to misinterpret the equals (=) sign. Because of the way it is commonly used in such expressions as $5+3=\_$, children tend to interpret = as an "operator" symbol—as meaning "adds up to" or "produces" (Baroody & Ginsburg, 1983; Behr, Erlwanger, & Nichols, 1980). Actually, = denotes an equivalence relationship: "the same as." Thus $8=5+3$ (or $8=8$) is mathematically correct because it means that 8 is the same as 5 plus 3 (or 8). But because children interpret = as an

TABLE 9.1: Facilitating a Broader View of Subtraction

All of these story situations can be solved by subtraction.

Name	Story Problem	Picture	Symbols
Take away	Cindy has 7 stickers. She gives 4 to Alison. How many does she have left?	Cindy: O O O O O O O Ø Ø Ø Ø O O O	7 - 4 = ? or 7 −4
Compare	Cindy has 7 stickers. Alison has 4. How many more stickers does Cindy have?	Cindy: O O O O O O O Alison: O O O O	7 - 4 = ? or 7 −4
Equalize	Cindy has 7 stickers. Alison has 4 stickers. How many more stickers does Alison have to get to have the same as Cindy?	Cindy: O O O O O O O Alison: O O O O	7 - 4 = ? or 7 −4

Based on Fuson (in press).

operator rather than as an equivalence symbol, they tend to think that such expressions are not acceptable. As a first grader put it: "[Equals] means it would add up to, and whatever the answer was, you'd put it down." This restricted view of "equals" may continue into high school and college and may affect mathematics learning at these levels (see, e.g., Byers & Herscovics, 1977). For instance, if "equals" is not viewed as a relational sign—as a bridge between numerically equivalent expressions—algebra solution strategies (such as adding identical elements to each side of an equation to simplify the expression on one side) may not be meaningful and may simply be learned by rote (Byers & Herscovics, 1977).

Both informal and formal experiences encourage the conception of "equals" as an operator rather than a relational symbol. Because of their informal adding experience, children come to school with a tendency to learn (assimilate) the symbol as meaning "adds up to." Early mathematics instruction then reinforces this interpretation of the symbol. Children are usually introduced to "equals" in the context of adding in the format of $1 + 1 = \_\_$. Workbook and ditto exercises reinforce this format, and the child becomes accustomed to "equals" implying "adds up to" (cf. Van de

Walle, 1980). Even the use of calculators encourages this notion: The arithmetic problem is punched in first, and then the equals sign key is hit to produce the answer. Children may reject or have problems with atypical forms, such as $\_ = 5 + 3$ or $5 + 3 = 3 + 5$, because they are generally unfamiliar with them (see Weaver, 1973).

A correct interpretation can be cultivated. The Wynroth curriculum (1969/1980) makes a concerted effort to teach a relational view of equals. "Equals" is defined as "the same as" to avoid the initial learning of "equals" as "the answer is." Moreover, the term "equals" is introduced not in the context of addition but in a manner that emphasizes a relational meaning. Dice games are used to teach the concept of addition. The child rolls the dice and the curriculum manual instructs the teacher to say, for example, "How much is 3 plus 4?" If necessary, the teacher might add, "What number did you get when you counted the dice—3 [point] plus 4 [point]?" No mention of "equals" is supposed to be made in this context.

Children first see "equals" in number sentences in the missing-number subsequence in such games as "Supposed to Be." In this game each player picks four number squares on which there is printed a one-digit number 0 to 9. The first player then draws a card on which there is an equation with a missing element (e.g., $3 + 2 = \_$). If the first player has a number square that would correctly fill in the missing element, that player may keep the equation card and draw a replacement square. The second player then draws a new equation card, and play continues. If the first player does not have a number square that correctly fits into the equation sentences, then the equation card is passed to the second player, who determines if any of his or her squares fit the equation sentence, and so forth. The first player, meanwhile, has the option of trading in one number square and picking a new one. The player who collects the most completed equation cards wins. It is important to note that this game immediately introduces children to a variety of equation forms ($1 + \_ = 3$, $\_ = 1 + 1$, etc.). Therefore, the child sees the equals sign in several contexts in an attempt to encourage a relational view of "equals."

Finally, the first written work (worksheets) involving equations is in the form of $4 = 4$, $4 \_ 4$, $4 = \_$, $5 \neq 3$, and so forth. Moreover, written addition is introduced in the form:

$$\begin{array}{c} 3 \\ 2 + 1 \end{array}$$

where the child writes the answer above the plus sign. The equals sign is first introduced with addition in the form $3 + 1 = 2 + 2$, $3 + 1 \neq 4 + 2$, or $3 + 1 < 4 + 3$, where the child would be asked to fill in a missing addend or relations sign (e.g., $3 + \_ = 2 + 2$ or $3 + 1 \_ 4 + 2$). Thus, the curriculum makes a concerted effort to encourage a relational rather than operator view of "equals."

This type of program has been fairly successful in cultivating a basic relational concept of "equals" (Baroody & Ginsburg, 1983). Children who had been in this program were more likely to see atypical forms as sensible and to give reasonable explanations for their judgments. Of particular interest was the children's reaction to atypical forms to which they had not been introduced. In fact, many children appeared to transfer their learning to these new cases. For instance, roman numerals had not been taught yet in any of the classes. Nevertheless, a number of participants said $7 + 6 = \text{XIII}$ made sense *if* XIII meant 13. A third-grade pupil's response illustrates this inferential process: "I don't know what this means. This is a 3 [pointing to III]; this—oh! This is 13 [indicating XIII] and 7 + 6 is 13, so that's right!"

Another example was provided by a first grade student's response to $7 + 6 = 6 + 6 + 1$, which she had not previously been exposed to. Asked if the math sentence was written correctly, she mentally computed the sum of $7 + 6$ and commented, "13 = 6 + 6 + 1 [laughs]. I think it [is] wrong. . . ."

INTERVIEWER: Is it written correctly?
SHERRY: Yes.
INTERVIEWER: Have you seen math sentences like this in math class?
SHERRY: I've seen stuff like that [7 + 6], but like that [6 + 6 + 1]?
INTERVIEWER: Nothing with three numbers?
SHERRY: No.
INTERVIEWER: What does 6 + 6 + 1 make?
SHERRY: That's 66.
INTERVIEWER: I think this is 6, + 6 + 1.
SHERRY: That's wrong. That's right!
INTERVIEWER: It's right?
SHERRY: Ahuh. 6 + 6 is 12, another 1 is 13.
INTERVIEWER: So, did Cookie Monster get this right or wrong?
SHERRY: Right. [Claps.] That's good.

However, even when instruction encourages a relational rather than an operational view of equals, teachers can expect children to cling to an "adds-up-to" view. The difficulty of learning a relational concept that conflicts with their informal operational definition is illustrated by the following exchange with a second grader:

INTERVIEWER: What does that say?

STUART: 13 [laughs] = 7 + 6.

INTERVIEWER: Did Cookie Monster write that correctly—like in math class?

STUART: You're supposed to start with this [7 + 6].

INTERVIEWER: Is it correct—does it make sense?

STUART: Yes, sometimes our teacher writes it backwards on the math worksheets.

INTERVIEWER: Should we put it in the right or wrong pile?

STUART: Both.

INTERVIEWER: If we can only put it in one pile. Does this make sense—to say 13 = 7 + 6?

STUART: No.

INTERVIEWER: Right or wrong pile?

STUART: Correct.

Teachers may not be able to eliminate but they can minimize children's cognitive resistance to learning an accurate meaning of equals by ensuring that instruction does not reinforce an operational meaning and carefully introducing a relational definition. This can be accomplished by introducing the equals sign in a context other than addition, as is done in the Wynroth curriculum. Specifically, the "equals" and nonequals signs could first be used to identify equivalent and nonequivalent sets of objects (e.g., • • • = • • • and • • • ≠ • • • •, respectively) and then with numerals (e.g., 3 = 3 or 3 ≠ 4, respectively).

Other Names for a Number

Children's informal concept of number tends to restrict the formal symbols a child identifies with a number. Children tend to equate the numeral 7 with the number "seven"—as "seven units." In fact, the numeral 7 is only one label or name for the number seven, which also goes by the names of 6 + 1, 9 – 2, 7 – 0, and so forth.

The School Mathematics Study Group (SMSG) advocates a name-for-number approach: teaching children that a number actually goes by a variety of names. This approach has been criticized because it requires understanding a relational definition of "equals," which is deemed unlearnable by children (e.g., Kieran, 1980). As noted above, careful teaching can encourage a relational view of "equals" in young children. Hence, the name-for-number approach may not be unwarranted. Introducing children to both typical and atypical addition equations would help children to see that there are various names for a number.

SUMMARY

Meaningful learning of skills depends on learning concepts and connecting symbols or procedures to concepts. Children's informal mathematics can provide a rich source of mathematical understanding. Teachers can make formal instruction more meaningful by connecting written symbols or definitions to children's informal concepts. Inadequate informal knowledge and gaps between informal knowledge and formal instruction are primary reasons for learning difficulties in typical and atypical children alike. Thus it is essential for educators to know how to cultivate and build upon children's informal knowledge. It is important to provide children the opportunities to discover key relationships informally. Children should be helped to see that formal mathematics is, in many cases, a way of representing what they already know. To foster meaningful learning, instruction should be sequenced in a manner that takes into account children's practical knowledge. Educators also need to be aware of how children's informal knowledge may limit or hinder their accurate learning of formal mathematics. Instead of working against the process of assimilation, teachers might better find ways to help children supplement their limited views.

CHAPTER 10

Basic Number-Combination Mastery

How are basic number combinations learned, mentally represented, and efficiently generated? Why is mastering the combinations so often such a difficult and prolonged process? What role do practice, meaningful learning, and informal methods play in mastering the basic number combinations?

NUMBER-COMBINATION DIFFICULTIES

Adam's Arithmetic

Adam automatically responded to the "double" $5 + 5$ because of his familiarity with money and to the double $8 + 8$ because of a nursery rhyme sung by his mother. Otherwise, the 11-year-old learning-disabled boy knew few basic addition or subtraction combinations, including zero combinations. For example, he often responded to problems such as $8 - 0 = \_$ with 0. Instead, he relied on informal computing: For instance, given $9 - 5 = \_$, he used a separating-from strategy with marks (ƚƚƚƚƚIIII).

Why had Adam not mastered the basic number combinations even after many years of remedial work? Was he incapable of learning number facts? Had he developed an overreliance on informal methods early in his school career? What could be done to help him master the basic number combinations?

A Key but Elusive Objective

Mastery of the basic number combinations is a major objective of primary-level mathematics education. Typically it is not an objective that is easily and quickly attained. Many educators find it difficult to get students to stop counting and to rely on recall. Student inability to learn number combinations is a common frustration among primary and special education teachers. In fact, one of the most common and significant deficits displayed by children having difficulty with math — including learning-disabled children — is a weakness in number-combination knowledge. Why is mastering the basic number combinations such a time-consuming and difficult task for many children?

TWO VIEWS OF MASTERING BASIC NUMBER COMBINATIONS

Absorption Model

Children just beginning school have few, if any, facts stored in long-term memory (LTM). Absorption theory posits that mastering the basic number facts is simply a matter of forming and strengthening *each* problem-answer association by means of practice (e.g., Ashcraft, 1985; Siegler & Shrager, 1984). For example, by repeated exposure, the fact "6 + 2 = 8" is ingrained in LTM (Thorndike, 1922). In a manner of speaking, children fill in a mental arithmetic table (Ashcraft, 1982). A completed mental addition table, depicted in Table 10.1, contains 100 separate facts. When given an addition problem, such as 2 + 6 = __, a "competent" child merely looks up the associated sum in the mental arithmetic table — retrieves the correct sum from LTM.

Absorption theory assumes that understanding is not necessary for learning the basic number combinations (Brownell, 1935). Because meaningful knowledge like commutativity is not considered a factor, the model implies that commuted pairs, such as 2 + 6 = 8 and 6 + 2 = 8, are learned separately. Specifically, the model proposes that if practiced less frequently, 2 + 6 = 8 will not be learned as quickly as 6 + 2 = 8. Moreover, the fact-retrieval system is seen as an autonomous unit that does not directly involve other arithmetic knowledge (Campbell & Graham, 1985). In Table 10.1, note that 2 + 6 = 8 and 6 + 2 = 8 are stored as separate facts in different locations in LTM. Therefore, recall of 2 + 6 = 8 and 6 + 2 = 8 are psychologically unrelated events. That is, the efficient production of 2 + 6 = 8 is unaffected by knowledge that 6 + 2 = 8 and knowledge of commutativity. In brief, mastering the basic combination simply entails rote memorization; efficient use involves only recall of specific facts from a network of facts.

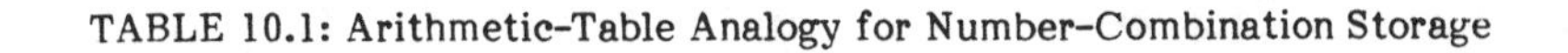

TABLE 10.1: Arithmetic-Table Analogy for Number-Combination Storage

Second Addend \ First Addend	0	1	2	3	4	5	6	7	8	9
0	0	1	2	3	4	5	6	7	8	9
1	1	2	3	4	5	6	7	8	9	10
2	2	3	4	5	6	7	8	9	10	11
3	3	4	5	6	7	8	9	10	11	12
4	4	5	6	7	8	9	10	11	12	13
5	5	6	7	8	9	10	11	12	13	14
6	6	7	8	9	10	11	12	13	14	15
7	7	8	9	10	11	12	13	14	15	16
8	8	9	10	11	12	13	14	15	16	17
9	9	10	11	12	13	14	15	16	17	18

A Cognitive View

In contrast to the absorption model, a cognitive model (Baroody, 1983b, 1984e, 1985a, 1985b; Baroody & Ginsburg, in press) suggests that general arithmetic knowledge (knowledge of rules, principles, and concepts) is an integral aspect of the learning, representation, and efficient production of the basic number combinations. This model posits that children do not learn the basic number combinations separately as specific numerical associations (as hundreds of feats of memory) but as a system of interrelated experiences (Carpenter, 1985; Olander, 1931). Moreover, it suggests that the mental representation of the basic number combinations entails relationships as well as facts. Thus the efficient production of basic combinations involves automatic thinking processes as well as the retrieval of data.

LEARNING RELATIONSHIPS. Mastering the basic number combinations may be largely a process of internalizing relationships rather than memorizing specific facts. Learning relationships can provide the basis for efficiently mastering whole families of basic combinations and related combinations. Take, for example, the zero rule that underlies the zero-addition family. Kindergarten children who did not know zero facts were given informal arithmetic experience that entailed providing concrete representations along with symbolic problems (Baroody, 1985b). For instance, a child was given a card with the written problem $4 + 0$ below a box with four dots and a box with no dots. To find the sum the child had to count the dots. Later the children were shown concrete counting all with blocks and fingers. After 8 weeks of training, the children responded quickly and correctly to nonpracticed zero problems and novel three-addend addition problems (e.g., $3 + 0 + 0$ or $0 + 0 + 2$), as well as practiced zero problems. From their informal arithmetic experience, the children abstracted a general rule or prescription that enabled them to respond efficiently even to previously unencountered zero problems.

The number combinations are replete with relationships that may be learned and serve as the basis for mastering basic combinations (Folsom, 1975; Jerman, 1970; Rathmell, 1978; Trivett, 1980). For example, subtraction is governed by an identity principle: $N - 0 = N$ (e.g., $9 - 0 = 9$, $19 - 0 = 19$). Children may also quickly learn an $N - N = 0$ rule to efficiently deal with problems such as $2 - 2 =$ __, $9 - 9 =$ __, or even $86 - 86 =$ __. For problems with terms that are number neighbors (e.g., $6 - 5 =$ __, $7 - 6 =$ __, $8 - 7 =$ __, or even $106 - 105 =$ __), a child might realize that the answer is always one. That is, the child may abstract the relationship: "The subtraction of 'number neighbors' produces a difference of one." Some subtraction combinations may be efficiently reconstructed from addition counterparts (e.g., $10 - 7$ is 3 because $7 + 3$ is 10) (Baroody et al., 1983; Carpenter & Moser, 1984; Steinberg, 1985).

STORING AND USING RELATIONSHIPS. It may be that internalized relationships are a key aspect of the mental representation and efficient production of number combinations. A quick and accurate response does not necessarily imply factual knowledge (see Example 10.1). For example, if an adult responds quickly with "one" to the problem $6 - 5 =$ __ (or $66 - 65 =$ __, $106 - 105 =$ __, or $766 - 765 =$ __), it does not necessarily

EXAMPLE 10.1

A Special Case of Mental Figuring

Some "idiot savants," mentally handicapped individuals with a special gift, have the uncanny ability to take a date, such as September 17, 1987, and quickly indicate the day of the week the date falls on: Thursday. It was long thought that a powerful rote-memory ability was the basis of such feats. That is, it was thought that calendars were memorized in the form of thousands of associations between dates and days or mental images of the calendars. Thus when given a date, all such a special individual had to do was look up in memory the correct fact (date-day association). However, such individuals can give correct answers to problems involving calendars they have never been exposed to (e.g., future dates). Thus, it is now believed that such problems are solved not by looking up stored facts but by quickly computing the answer. What is stored in memory appears to be an algorithm—a computational procedure—for figuring out date-day connections. In short, it seems that these special individuals use a reasoning rather than a recall process to efficiently generate calendar solutions.

mean the person has stored in and recalled from LTM the fact "6 – 5 = 1" (or "66 – 65 = 1," "106 – 105 = 1," or "766 – 765 = 1"). It is quite possible that the adult continues to use the difference-of-one relationship learned as a child. Indeed, well-learned rules, principles, and thinking strategies (prescriptions) may account for the efficient production of many number combinations.

INTEGRATED KNOWLEDGE. In this view, mastering, representing, and processing basic combinations is intimately linked with general arithmetic knowledge. General arithmetic knowledge provides the framework for organizing and learning an otherwise large array of new information. For example, it is not necessary to memorize each of the 100 single-digit addition combinations because, except for the 10 "doubles" (e.g., 2 + 2, 5 + 5, 9 + 9), half are related to the other half by the principle of commutativity (e.g., Folsom, 1975). Thus knowledge of this principle, which is acquired rather early (Baroody & Gannon, 1984; Baroody et al., 1983), might enable a child to learn both 2 + 6 = 8 and 6 + 2 = 8, even though the first combination is practiced considerably less than the second. Prescriptions, such as commutativity, make it unnecessary to represent and process every possible number combination separately.

Unlike absorption theory, then, this cognitive model proposes that number combination and meaningful arithmetic learning go hand in hand. Of course, children can and do memorize isolated facts that have little meaning to them. But for the most part, children's "number-fact" facility should grow as their understanding of arithmetic evolves. As general mathematical knowledge becomes more extensive, integrated, and automatic, number-combination facility should grow correspondingly (Baroody & Ginsburg, in press). As children internalize mathematical relationships, they have a broader base for learning the basic number combinations.

EDUCATIONAL IMPLICATIONS: DEVELOPMENTAL FACTORS AND PITFALLS

The Role of Informal Experience

According to some proponents of absorption theory, informal methods, such as counting or thinking strategies, are not central to the task of memorizing the basic number combinations. Indeed, counting and thinking strategies are sometimes viewed as hindrances—as attempts to evade the real work of rotely learning the number facts. For example, Wheeler (1939, p. 311) commented: "We wonder if the children are not computing the sums by physical or mental *counting*, a *crutch* which is probably developed in the child while building the number concepts" [emphasis added]. Similarly, Smith (1921, pp. 764–65) considered thinking strategies an impediment to learning the facts:

> Another pupil required a long time for the sum of 6 and 9. He explained his process as follows, "6 and 10 are 16; 6 and 9 are 1 less than 6 and 10; then 6 and 9 are 15." He had to think through a similar form every time any number was added to 9 and of course gave much slower responses. . . . We should be careful about letting pupils acquire forms or roundabout schemes for securing a result in the lower grades which will prove a handicap to them in the upper grades.

Unlike these early accounts of absorption theory, a cognitive model suggests that a period of informal calculating and exploration is important for number-combination mastery.[1] Counting experience may permit children to discover mathematical relationships that can serve as prescriptions for number combinations. For example, by calculating the differences for a

1. Psychologists now generally agree that a period of informal calculating is important for learning the basic number combinations. Modern absorption theories (e.g., Ashcraft, 1982; Siegler & Shrager, 1985) note that informal computational practice is an important means of strengthening an association between problem and answer. Even so, modern absorption theories overlook other important ways informal experience contributes to mastery.

sequence of problems like 2 − 1 = __, 3 − 2 = __, 4 − 3 = __, 5 − 4 = __, a child might notice the difference-of-one relationship.

The Role of Practice

According to absorption theory, practice is the determining factor in number-combination mastery (e.g., Ashcraft, 1985). Every time a child repeats a combination a trace is left in LTM (Siegler & Shrager, 1984). In time, the trace strength builds up, and the association between the correct answer and a problem is firmly established.

The cognitive model suggests that the amount of practice may not be *the* determining factor in mastering the basic number combinations. In fact, empirical evidence (e.g., Olander, 1931; Thiele, 1938; Wheeler, 1939) indicates that the amount of practice is not predictive of mastery. It may be that a sum is computed repeatedly with little or no impact on LTM. Like an unimportant telephone number that is needed only for the moment, the problem and computed sum are held in working memory but not entered into LTM.

Sometimes, though, as with a telephone number that forms an interesting and easily definable pattern, a combination may make a mark on LTM without conscious effort. The small addition doubles (1 + 1, 2 + 2, 3 + 3, 4 + 4, 5 + 5) may be particularly susceptible to incidental learning because of the abundance of readily recognizable patterns. Many objects in the child's environment have a symmetrical number of parts that lend themselves readily to a mental image. For example, the human body has one eye on each side of the head for a total of two eyes. The hands provide a memorable model for 5 + 5 = 10. Moreover, the other doubles are relatively easy to represent on the fingers: The child simply has to put up the same number of fingers on each hand. This may give the small double a special status in the child's mind. Numerous objects in the environment (e.g., two wheels on each side of a car makes four) may provide models for the doubles. With dice, the doubles form distinct patterns that may provide the basis for a mental image of the part-part-whole relationships of the doubles. Not surprisingly, the small addition doubles are among the earliest number combinations mastered.

As with important telephone numbers, children may consciously make an effort to memorize some combinations. Interest, which may be motivated by a need for approval or a fear of disapproval, may provide a powerful incentive to learn combinations. Very few repetitions may be required if a child has a great interest in mastering a combination (Thorndike, 1922). The alternative model suggests, then, that practice does not automatically leave traces (build up associative strengths) in LTM.

Is practice or drill unimportant? Practice is important because it can serve to make the use of rules, principles, and thinking strategies, as well as specific facts, automatic. Well-rehearsed knowledge can lead to the discovery of new relationships. The child can also use prescriptions and facts to process harder problems. For example, mastery of many basic subtraction combinations may depend on first mastering basic addition combinations (Baroody et al., 1983). Addition and subtraction are related to each other by the complement principle: Subtracting a part (an addend) from a whole (the sum) yields the other part (addend) (e.g., $6 + 2$ or $2 + 6 = 8$ and $8 - 6 = 2$ or $8 - 2 = 6$). During their efforts to compute differences, children are more likely to discover this important relationship if they can readily call to mind the addition counterparts. For example, a child who can quickly remember that $6 + 2 = 8$ may connect this knowledge to the computed problem $8 - 6 = 2$. After the connection is discovered, automatic addition combinations, such as $6 + 2$ or $2 + 6 = 8$, can serve as the data for efficiently generating the answer to subtraction problems, such as $8 - 6 = \_$ and $8 - 2 = \_$.

Combination Difficulty

Absorption theory implies that because it simply involves a rote process, mastery of the basic number combinations should be quick and rather uniform. That is, given adequate practice, all the basic number combinations should be mastered in the first few years of school. Moreover, given an equal amount of practice, large problems such as the nines should be no more difficult to master than small problems such as the twos. That is, "psychologically the child should be able to learn $5 + 4 = 9$ as easily as $2 + 3 = 5$" (Wheeler, 1939). In effect, the rate and order of mastering the basic number facts is determined by how frequently each combination is practiced.

The cognitive model suggests that the rate and order of number-combination mastery is more directly linked to the development of meaningful knowledge than to practice frequency (Baroody & Ginsburg, in press). Because it takes time to "see," assimilate, and build up a network of relationships, mastery of the number combinations is necessarily a gradual process. Moreover, whereas some relationships are relatively easy to see and are quickly internalized, others are not easily abstracted and require time to master.

ZERO COMBINATIONS ($N + 0$ and $0 + N$, $N - 0$, $N - N$, and $N \times 0$ and $0 \times N$). The $N + 0$ and $0 + N$ combination family can be generated by the easily internalized and implemented identity ($N + 0$ or $0 + N = N$) relationship (Baroody, 1985a). This may explain why the combinations involving zero are the most rapidly generated addition combinations (re: Svenson, 1975, Figure 2) — even among children just beginning school (re: Groen & Parkman, 1972, Figure 2). Likewise, the $N \times 0$ ($0 \times N$), $N - 0$, and $N - N$ combinations can be generated by relatively discernible and straightforward relationships ($N \times 0$ or $0 \times N = 0$, $N - 0 = N$, and $N - N = 0$, respectively). This may explain, for example, the relative quickness of the $N - 0$ and $N - N$ subtraction combinations as early as the second grade (Woods et al., 1975). Because the zero rules are so similar, children may occasionally make slips such as saying $7 + 0$ is 0 or 7×0 is 7.

ONE COMBINATIONS ($N + 1$ and $1 + N$, $N - 1$, and $N \times 1$ and $1 \times N$). The one combinations are among the earliest mastered by children. For addition and subtraction, children may discover they can use the already well-learned number-after and number-before number-sequence relationships. The one-times combinations are based on an easily learned identity relationship ($N \times 1$ or $1 \times N = N$).

TWO ADDITION COMBINATIONS ($N + 2$ and $2 + N$). Though not as easily mastered as the previous families, this family is easier than those listed later. The relative ease of the plus-two/two-plus family may be due — like the $N + 1$ and $1 + N$ family — to children's counting facility. The initial counting-on procedure for dealing with these combinations is relatively easy and, with a modicum of practice, can be done quite quickly. Children may then increase production speed by using their mental number list: the "skip-next-N" rule (for $N + 2$ and $2 + N$ combinations, skip over the number after N).

LARGE ADDITION DOUBLES. Though more difficult to master than the small doubles and many other small number combinations, the large doubles are among the earliest of the larger combinations to become automatic. The combination $6 + 6$ probably becomes quite familiar to children accustomed to playing games with dice. Combinations such as $7 + 7$, $8 + 8$, and $9 + 9$ may be memorized as specific facts.

ADDITION DOUBLES + 1. Once children have learned the doubles, they often discover the doubles-plus-one (-minus-one) relationship (e.g., $5 + 4$ is one more than $4 + 4$ [or one less than $5 + 5$] and so the sum must be nine). In time this thinking strategy can become quite automatic.

SUMS EQUAL TEN. This family may be unique in that the sum corresponds to children's endowment of fingers. Children learn relatively early how many fingers they have and that ten is a special number. Moreover, the addends in this family can be represented on the fingers either overtly (finger counting) or covertly (kinesthetically). With some practice, the child may learn to recognize that certain combinations of fingers (e.g., $7 + 3$, $8 + 2$) produce the special case: ten (a sum involving all fingers).

NINES ($9 + N$ and $N + 9$). Mastery of the nines often involves building on easier combinations and thus is usually delayed. These combinations are often dealt with by means of reorganizing the addends to form a problem involving ten (e.g., $9 + 7 = [9 + 1] + [7 - 1] = 10 + 6 = 16$). This approach depends on a relatively well-developed number sense. Clearly, for this redistribution process to become automatic, ten combinations must be automatic. This redistribution process can be short cut with the "Nine Rule": "The sum of $N + 9$ or $9 + N$ is the number before $N +$ *-teen" (e.g.,* $7 + 9$ is the number before seven — six — plus -teen). Some adults continue to use such thinking strategies to quickly reason out nine combinations (Browne, 1906; Parkman & Groen, 1971).

REDISTRIBUTION BASED ON TEN. The redistribution strategy that can be applied to $9 + N$ and $N + 9$ combinations can be applied to combinations such as $7 + 4$, $7 + 5$, $8 + 4$, $8 + 5$, $8 + 6$: Decompose the smaller addend in such a way as to make the larger addend ten, and add the remainder to ten (e.g., $6 + 8$: $[6 - 2] + [8 + 2] = 4 + 10 = 14$). Because a number greater than one must be exchanged, this thinking strategy is more difficult than redistribution involving nine.

DIFFERENCE OF ONE. Problems such as $3 - 2$ or $9 - 8$ form a special family of combinations because they can be based on a simple but not obvious relationship ("The difference of two number neighbors is one") and the well-learned number sequence.

ADDITION COMPLEMENTS. Mastery of many subtraction combinations (e.g., $5 - 3$, $7 - 4$, $9 - 5$) may be delayed until after a child discovers the relatively subtle addition-subtraction complement principles (e.g., $2 + 3$ or $3 + 2 = 5$ and so $5 - 3 = 2$ and $5 - 2 = 3$) and masters the corresponding addition combinations (e.g., $2 + 3$ or $3 + 2 = 5$).

A Cognitive View of Difficulties

The cognitive model helps to account for the weakness in basic number-combination knowledge that is characteristic of children having difficulty with mathematics (Allardice & Ginsburg, 1983; Baroody, 1983a; Russell & Ginsburg, 1984) and why larger problems (e.g., Kraner, 1980; Smith, 1921) are especially difficult for such children. Difficulty with such combinations by "learning-disabled" children may be due not to a memory deficit or inadequate practice but rather to the lack of a rich network of relationships (Baroody, 1985a). That is, they may not have a rich network of rules, principles, and thinking strategies to use for producing larger number combinations. Thus, such a child is faced with the burden of memorizing many apparently isolated facts. "Learning-disabled" children may feel overwhelmed with such a chore and may give up learning the combinations.

For example, Adam—even after six years of special education—did not know zero combinations like $5 - 0 = 5$ and $8 - 0 = 8$. Despite years of practice, he had failed to abstract an $N - 0 = N$ relationship. If this regularity was pointed out, he would quickly respond to an $N - 0 = \_$ problem, but the improvement was only temporary. Though he could use an $N - 0 = N$ rule appropriately, it was quickly forgotten. A year after I first saw him, Adam knew $N - 0 = N$ combinations. It may be that he could not retain the relationship until he saw a meaningful connection—understood an identity principle.

Two interrelated factors may have contributed to Adam's difficulties in mastering the basic subtraction combinations: his failure to learn the basic addition combinations, and his failure to learn the addition-subtraction complement principle. Adam did not adequately understand the complement principle because he had not mastered the basic addition combinations. If a child cannot quickly and accurately remember that $8 + 7 = 15$, then using the complement principle with $15 - 8$ or $15 - 7$ is not going to be efficient. Moreover, a child who does not mentally have the addition facts readily available is much less likely to discover the connection between addition and subtraction. (This is especially true for a child whose attention is absorbed in executing a less-than-automatic counting procedure for subtraction.) Thus a child like Adam may have little success "learning" the basic subtraction facts until the basic addition combinations and the complement principle are mastered.

EDUCATIONAL IMPLICATIONS: MEANINGFUL INSTRUCTION OF BASIC NUMBER COMBINATIONS

In contrast to the absorption model, the cognitive model discussed suggests that mastery of the basic number combinations is best achieved by fostering informal exploration and meaningful learning rather than rote memorization. General guidelines for achieving this aim are delineated below. Though the suggestions may take more time and teacher effort than direct-instruction-and-drill methods, children should retain and apply their basic arithmetic knowledge more effectively.

1. *Encourage the search for and discussion of relationships.* The discovery of relationships, such as "adding zero leaves a number unchanged" or "the difference of two number neighbors is one," can provide a basis for efficiently generating a family of basic number combinations as well as a whole range of related, multidigit combinations. By asking children to look for relationships and discuss the patterns they find, the process of meaningful learning and thus number-combination mastery is facilitated. In addition, such an approach underscores the real nature of mathematics, and it is an approach that children typically enjoy.

2. *Establish a firm foundation: efficient informal calculation.* Children need ample opportunity to compute arithmetic solutions in ways that are meaningful to them. Informal calculation can provide an important basis for discovering numerical relationships in a meaningful manner. Children are more likely to discover such relationships if their informal computing procedures are efficient. If computation is not consistently accurate, abstraction of relationships will be impeded; if it is not automatic, patterns may not be noticed because the effort of computing absorbs so much attention. Remedial efforts should be aimed at eliminating these deficiencies.

3. *Encourage, point out, and discuss thinking strategies.* Thinking strategies appear to be a key intermediate step between computing and mastery for many children (e.g., Brownell, 1935). Indeed, some research (e.g., Brownell & Chazal, 1935; Steinberg, 1985; Swenson, 1949; Thiele, 1938; Thornton, 1978) indicates that teaching children "thinking strategies" is more effective than drill in facilitating the retention and transfer of basic number combinations (Carpenter, 1985; Suydam & Weaver, 1975). Such an approach may be especially important for and can be effective with learning-disabled children (Thornton & Toohey, 1985). Pointing out and discussing thinking strategies has a number of important consequences. (a) It may help to make clear and facilitate the learning of relationships that may have otherwise gone unnoticed by a child. (b) Children are less likely to equate mathematics with memorizing facts and more likely to equate it with the discovery and application of regularities. (c) It validates the child's informal mathematics. In effect, the teacher endorses how the child thinks mathematically, and a child may discover that others think as he or she does. (Adults are sometimes surprised and relieved that some other adults still use thinking strategies.) (d) It engenders interest and motivation to learn. Children are sometimes surprised and usually delighted when shown that they can reason out answers. (e) General learning and problem-solving skills may be enhanced. For example, children may be more likely to search for patterns in, say, spelling.

4. *Use drill effectively.* Practice can make use of prescriptions (rules,

principles, and thinking strategies) automatic. It can also foster the internalization of specific facts. Drill is an important component of instruction after children have had the opportunity to learn relationships.

Drill need not be dull. Indeed, ample practice can be provided by playing games such as those listed below. Practice in an interesting context will promote positive rather than negative attitudes toward mathematics.

Games and Activities

DICE BASEBALL

Objective: Practice basic addition combinations 1 + 1 to 6 + 6

Materials: (1) Dice
(2) Board with baseball diamond
(3) Tokens to represent batter and runners

Instructions: Two teams of one or more children can play. The child "at bat" throws the dice. After determining the value (sum) of the throw, the outcome is determined by the key below:

THROW = OUTCOME

2 = triple (batter advances three bases)
3 = out
4 = walk
5 = out
6 = out
7 = out
8 = single (batter advances to first base and any runners advance one base
9 = out
10 = double (batter advances to second base and any runners advance two bases)
11 = out
12 = home run

Have the children keep score and keep track of the number of outs. After three outs, the "side" is retired, and the other team takes their turn at bat.

CLEAR THE MAZE

Objective: Addition (or other math skills in which the student gives a small whole-number answer)

Materials: (1) Gameboard, divided in half, with a maze on each side
(2) Small chips or blocks
(3) Addition problem cards or dice

Instructions: Put a chip on every square of both mazes. Tell the players: "We're going to see who can get rid of all the chips on their maze first. The cards (dice) will tell us how many we get to take off." After figuring out the addition problem, the player removes that number of chips from his or her maze.

BREAKDOWN

Objective: Practice basic arithmetic combinations
Materials: (1) Set of cards with arithmetic facts to be practiced
(2) Blocks (to compute sums, as needed)
(3) Stopwatch or other timing device
Instructions: Lay the cards face down on the table in the form of a brick wall. The child turns the cards over one at a time and supplies an answer to the problem posed. If the answer is correct, the child removes the card from the wall. If the answer is incorrect, the player may compute the sum with objects (e.g., blocks or fingers) and then remove the card. This is repeated until the entire wall is gone. Note the time required to break down the wall. In successive games, encourage the child to break his or her previous time record.

WHAT ADDS UP TO?

Objectives: (1) Name-for-a-number concept (a number has various names–e.g., 7 is but one name for a number that can also be named $7-0$, $6+1$, $9-2$)
(2) Practice using patterns and principles
(3) Practice number combinations
Materials: Paper and pencil
Instructions: Explain, "When I say, 'Go,' write down as many arithmetic combinations with answers of 14 as you can." A sample response is shown below.

$14+0$	$14-0$
$13+1$	$15-1$
$12+2$	$16-2$
$11+3$	$17-3$
$10+4$	14×1
$7+7$	7×2

The activity can be restricted to a single operation (e.g., "Write down as many addition combinations with *sums* of 14 as you can"). To encourage efficiency, a time limit can be set for advanced players.

To help children recognize the utility of arithmetic principles and number-combination patterns, explicitly encourage them to exploit relationships or regularities. Children can work on the exercise singly or in small discussion groups. This should be followed by a teacher-led discussion of the exercise. An alternative approach is to give the students a sample response like the one shown above and ask them to note what relationships or regularities the "sample child" used or failed to use. For example, the main points to cover in analyzing the above sample would be:

- The series of the first five combinations could have been expanded further to include $9 + 5$ and $8 + 6$. In other words, the child could have used a simple compensation (take one away from the first addend and add it to the second) to create a whole series of answers. Similarly, the sequence of subtraction combinations could have been extended by incrementing both the minuend and the subtrahend: $18 - 4$, $19 - 5$, $20 - 6$, and so forth.
- The sample child used the identity principle for both addition and subtraction ($14 + 0$ and $14 - 0$).
- Base-ten patterns were not fully exploited. Though the child did give $10 + 4$ as an answer, many other combinations involving 10 and multiples of 10 are available, such as $24 - 10$, $34 - 20$, and so forth.
- The commutativity principle was not exploited for addition and multiplication. Applying this principle would have nearly doubled the sample child's output for addition and would have doubled it for multiplication.

SUMMARY

Children appear to learn the basic number combinations in large part as a system of relationships, not as separate associations (as hundreds of isolated facts). Prescriptions (principles, rules, and thinking strategies), which are integral to a child's general knowledge of arithmetic, may underlie the efficient production of many combinations. Meaningful and informal learning then are key to mastering the basic number combinations. Because there is no quick and easy way of promoting such learning, educators should expect mastery to be a prolonged process. Meaningful instruction is more likely to encourage the mastery of the basic number combinations than a drill approach is. Guided discovery of relationships, informal computing, and thinking strategies are key ingredients for fostering mastery. Practice is important but should follow meaningful learning and be made interesting.

CHAPTER 11

Reading and Writing Basic Symbols

What accounts for common difficulties like reading 9 as "six" or writing numerals in reverse? Are such problems typically due to inadequate practice or a perceptual-motor difficulty? What kind of training program is most likely to avoid or remedy such problems?

READING AND WRITING DIFFICULTIES

The Case of LeRoy

LeRoy, a mildly mentally handicapped inner-city boy of 9 years,[1] read most of the single-digit numerals except for 2, which he identified as "four," and 9, which he called "six." Even with retesting he continued to read 9 as "six." As Figure 11.1 shows, he could successfully write only four numerals. LeRoy reversed 4, reversed and wrote a barely recognizable 5, and was altogether unsuccessful with 6, 8, and 9. Interestingly, before each attempt to write 6, 8, and 9, he asked, "How do you make it?" Even with a model of the numeral in front of him, LeRoy copied 4 and 9 in reverse and had great difficulty with 6. Throughout the testing LeRoy eagerly asked: "Is this a six?" "Is this a nine?"

Apparently LeRoy could not discriminate between 6s and 9s. He wrote some numerals in reverse, even when he had a model numeral in front of him to copy. Was such confusion due to an immature perceptual system that did not allow him to see written symbols accurately? That is, did he sometimes see objects in reverse, or did the signals sent to the brain sometimes get jumbled? What else could explain LeRoy's difficulties? His constant questioning suggested that he was interested in learning and that his difficulties could not be attributed to poor motivation.

1. This case study was conducted by Elizabeth K. Messner.

FIGURE 11.1: LeRoy's Numeral Writing and Copying

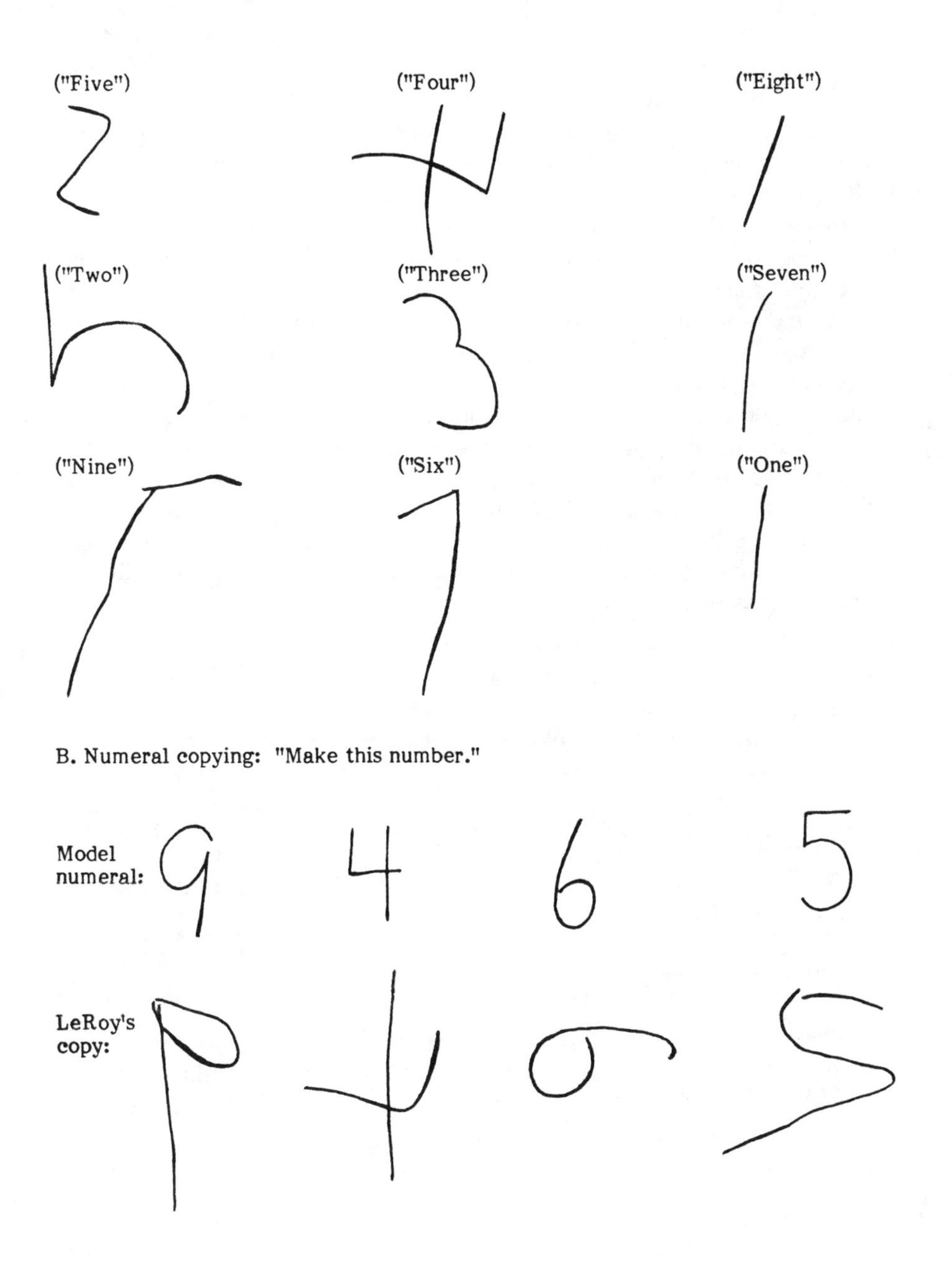

Knowledge of Written Symbols

School mathematics requires the use of written symbols to represent numbers (e.g., numerals such as 6, 24, 3.5, and 1/2), arithmetic operations (+, −, ×, ÷), and key mathematical relations: = (equals), ≠ (not equal to), > (greater than), and < (less than). Recognizing, reading, and writing the mathematical symbols is a fundamental objective of formal instruction. Symbol recognition (the ability to identify, say, by pointing out the symbol) is developmentally more basic than symbol reading: recognizing the symbol and saying its associated name (e.g., labeling the symbol = as "equals"). Writing a symbol is a relatively sophisticated skill because it involves both recognition and translating a mental picture of the symbol into motor actions.

Even though preschoolers may be able to count up to 20 or even beyond, they may be unfamiliar with written numbers. By the time they enter kindergarten, most children, though, can recognize and read one-digit numerals. By the end of kindergarten, most can also write one-digit terms. During the primary years, with the possible exception of the greater-than (>) and less-than (<) symbols, children typically master basic mathematical symbols without much difficulty. The same cannot be said of special education children. What accounts for the reading and writing difficulties prevalent among learning-disabled and mentally handicapped children? What kinds of remedial programs are commonly recommended, and which is most effective?

DIFFERENT VIEWS—DIFFERENT INSTRUCTION

Psychologists have adduced very different theories about how reading and writing skills develop and why children experience difficulties. These different theoretical perspectives suggest different instructional and remedial approaches.

Popular Instructional and Remedial Approaches

THE ABSORPTION MODEL. According to the absorption model, learning to recognize and read symbols basically entails forming an association between a symbol and its name. Learning to write symbols essentially requires establishing correct and automatic motor habits. Thus the emphasis of training is modeling, imitation, and practice. To achieve overlearning, remedial efforts with special education children consist of extensive practice.

Writing practice in many initial or remedial programs takes the form of tracing and copying (Fernald, 1943). For example, children are encouraged to trace their fingers along felt or sandpaper cutouts of numerals or make clay copies of numerals. Another common practice is to have children trace over dotted lines outlining the numerals (Engelhardt et al., 1984). The assumption is that the repetition of the motor pattern will eventually etch a mental image of the numerals into memory, enabling the child to recognize and read numerals. Moreover, the practice will also establish an automatic habit pattern that will enable the child to write the numerals.

THE PERCEPTUAL-MOTOR DEFICIT MODEL. A prevalent view, especially in the field of special education, is that reading and writing problems are often due to a more basic underlying problem. According to this view, reading or writing difficulties, especially among learning-disabled and mentally handicapped children, can be attributed to deficiencies in general perceptual-motor abilities.

Identification and reading problems have been attributed to general perceptual difficulties, such as poor figure-ground perception: an inability to distinguish a detail from its context (e.g., Bley & Thornton, 1981; Frostig, LeFever, & Whittlesey, 1964). Some theorists (e.g., Orton, 1937) suggest that numeral confusions, such as reading 9 as "six," are due to inadequate perceptual-motor activity that results in "incomplete dominance": the inability of the left or right hemisphere to control the interpretation of incoming signals. The implication is that sometimes the image in the left hemisphere is used to read a numeral and sometimes the image in the right hemisphere is used. It is assumed that because the image in one hemisphere is the inverse of that in the other, the child sometimes sees the 9 correctly and at other times sees its inverted image.

The perceptual-motor deficit model assumes that writing problems are due to general deficiencies in processing visual information, controlling hand movements, or coordinating visual and motor skills (e.g, Frostig et al., 1964). For example, copying numerals backward (e.g., ᒥ for 7, ∂ for 6) has been attributed to perceptual difficulties that preclude seeing accurately what is presented or visual-motor integration deficiencies that prevent the proper coordination of eyes and hand movements (Bley & Thornton, 1981). Improper numeral formation (e.g., ⊣ for 7) or difficulties in writing on lined paper (e.g., 7) have been ascribed to visual-spatial problems or fine-motor-coordination deficiencies (Bley & Thornton, 1981).

Proponents of a perceptual-motor deficit model (e.g., Barsch, 1965; Delacato, 1966) argue that remedial efforts must correct the underlying perceptual-motor difficulties before a reading or writing problem is addressed directly. In other words, if the primary cause of such academic difficulties is a perceptual-motor deficit, then remedial efforts must first focus on the development of general perceptual-motor skills. Thus children with reading or writing problems are sometimes given exercises from programs that are directed at fostering visual-motor and fine motor skills (e.g., Pittsburgh Perceptual Skills Curriculum, Rosner, 1971a, 1971b).

A Cognitive Model

RECOGNITION AND READING. A cognitive model (e.g., Gibson & Levin, 1975) proposes that to recognize or read a symbol, a child must know what distinguishes it from other symbols. For example, to differentiate among the numerals, a child must know the defining characteristics of each numeral: the component parts and the part-whole relationships (how the parts fit together to form a whole). Specifically, the defining characteristics of 6 include a line that curves out to the left and a single "loop" on the bottom right of the curved line. The curved line distinguishes a 6 from numerals that contain only straight lines (1, 4, and 7). The single loop distinguishes 6 from all other numerals except 9. The direction of the curve and the position and direction of the loop are the only features that separate 6s and 9s. Thus recognizing and reading numerals involves a sophisticated visual analysis that entails learning the distinguishing characteristics of each numeral.

RECOGNITION AND READING DIFFICULTIES. According to this cognitive model, most recognition and reading difficulties are due to not knowing the defining characteristics of symbols rather than to inadequate practice or general perceptual-motor deficiencies (e.g., Gibson & Levin, 1975). The more similar the forms (the more defining characteristics that are shared), the more likely that a child will confuse symbols. Confusion is especially likely where orientation is the key or only distinguishing feature. This explains why children have so much difficulty learning to recognize and read 6, 9, <, and >. *Even though children can see these symbols accurately*, they may have difficulty labeling the orientation (e.g., "right," "left") and connecting these directional labels to the correct symbol (e.g., the point of the greater-than sign points to the right).

WRITING NUMERALS. Learning to write numerals is initially more difficult than learning to read them because it requires a set of rules for translating an image (the defining characteristics) into motor actions (Kirk, 1981). The rules specify where to start and how to proceed (Goodnow & Levine, 1973). Thus, not only must children know what a numeral looks like, they must have a motor plan that directs the writing process. For example, to write a 7, children must know—ahead of time—that they should start at the upper left, draw a horizontal line from left to right, change direction, and make a diagonal that ends up in the lower left. In brief, numeral writing encompasses learning the defining characteristics and a step-by-step motor plan for recreating the component parts and part-whole relationships for each numeral.

WRITING DIFFICULTIES. Most writing difficulties are probably due to a deficient motor plan rather than to general perceptual-motor problems (Kirk, 1981). Consider the child who can recognize and read numerals but cannot write them (e.g., LeRoy could read but not write 8s). Such a child can see the numerals correctly, knows the distinguishing characteristics, and probably has a mental picture of the numerals. However, without a motor plan, the child cannot print the numerals. Moreover, consider the child who consistently writes a numeral backward and realizes there is a problem but cannot correct the error (see Example 11.1). To write numerals correctly, children must know before they start where to begin and in what direction to head. For example, in making a nine, the child must start in the upper right and proceed to the left (9). If the child has a left-to-right preference (a tendency encouraged by reading instruction), the child will repeatedly write 9s backward (9), even though he or she *sees* that it is incorrect.

Without a motor plan, even making a copy of a numeral is a difficult, if not impossible, task. LeRoy intuitively realized that he did not have a step-by-step plan for writing or copying all the numerals. (It was this knowledge that prompted his question "How do you make a nine?") Thus even with a model 9 present, he did know *how* to go about writing the numeral.

EDUCATIONAL IMPLICATIONS: FOCUSED READING AND WRITING TRAINING

In most cases, training efforts should focus directly on reading and writing subskills rather than on practice or general perceptual-motor skills. Specifically, training needs to make explicit the distinguishing characteristics of and the motor plan for symbols.

Modeling, imitation, and practice are important but are not the key elements in teaching reading and writing skills. Young children and spe-

EXAMPLE 11.1

The Case of Steven

Steven, an 11-year-old moderately retarded child, could already recognize and read the numerals 1 to 10. However, except for 1, 2, and 4, he was unable to write the single-digit numerals. One problem that plagued Steven was reversals. Asked to draw a seven, he drew ϒ and commented, "Oh, this is a 't,' this isn't right." For "three," he drew Ƽ and said, "This is a 's.' How do you make a three?" How would a perceptual-motor deficit and a cognitive model explain Steven's responses? The perceptual-motor deficit model would attribute Steven's writing problems to an inability to see the correct orientation of the numerals, form an accurate mental image, or control his fine motor actions. A cognitive theorist would point out that the child appears to be aware that he drew the seven and three incorrectly because the written forms do not match the child's (correct) mental image of the numerals. Though the child knows the defining characteristics of the numerals, he did not have a preplanned method for translating the mental image into appropriate motor actions. Indeed, like LeRoy, Steven was <u>aware</u> that he did not have a motor plan and quite sensibly asked the tester for guidance. Moreover, Steven had the fine motor coordination to draw fours, which is <u>more intricate</u> than drawing a seven.

This case study was conducted by William J. Boaz.

cial education students may need help making a visual analysis of the numerals so as to learn their defining characteristics. Such children often also need assistance in mastering motor plans for the numerals. Instruction or remediation that does not focus on these key elements may not be productive and may even be counterproductive. For example, though children may discover the defining characteristics of numerals or devise a motor plan during the course of tracing-and-copying exercises, practice in itself does not guarantee such learning. Thus a considerable amount of time and effort may be wasted in practice that does not help children to learn the key skills for reading and writing numerals. Moreover, it is very frustrating for children to continue practicing, say, writing numerals when they can see that they are doing it incorrectly but do not know how to correct their motor plan. Practice is important to make skills automatic once a child has learned the defining characteristics or a motor plan.

Perceptual-motor deficits appear to be greatly overrated as a cause of reading and writing problems (e.g., Vellutino, Steger, Moyer, Harding, & Niles, 1977). In most cases, remedial efforts that concentrate on improving general perceptual-motor skills will not redress the real problem. Even in cases where there is genuine perceptual-motor difficulty, perceptual-motor training and practice is probably not sufficient to correct a reading or writing deficiency. Though steps need to be taken in such cases to correct a child's vision, develop fine motor coordination, and so forth, remedial efforts must also focus directly on the academic skills. If training does not directly deal with learning the defining characteristics or motor plans, it probably will not be successful in helping children to read or write numerals. Indeed, research (e.g., see reviews by Gibson & Levin, 1975; Hallahan & Cruickshank, 1973; Vellutino et al., 1977) indicates that remedial approaches that emphasize general perceptual-motor training may promote the development of perceptual-motor skills but are not fruitful in correcting specific academic deficiencies.

1. *Recognition and reading training should focus on pointing out the defining characteristics.* For example, in making clay models of numerals, it is important for the teacher to point out the component parts of a numeral and how the component parts fit together. If a child has not mastered the vocabulary that describes defining characteristics (e.g., top, bottom, right, and left), this deficiency should be remedied first. To some extent, a vocabulary deficiency can be circumvented. For example, for right-handed children, "toward your pencil hand" can be substituted for "right" and "toward your free hand" for "left." Or easily identifiable pictures can be included on the writing paper to identify position (e.g., a star at the top of the page, a crescent moon at the bottom, a plane on the left-hand side of the page, and a car on the right).

Easily confused symbols should be taught together. For example, by placing 6 and 9 or 2 and 5 side by side and explicitly noting how they differ, the child has a better chance of learning the features that distinguish the numerals. Wynroth (1969/1980) suggests a very effective memory aid for keeping the greater-than and less-than symbols straight: "Cookie Monster's mouth is always open to the larger amount of cookies" (e.g., $5>3$: five cookies are more than three, $2<4$: four cookies are more than two or two cookies are less than four). Note that this aid (rule) does not depend on knowing the directions "right" and "left" or keeping straight an association between position and definition ("open to the left means 'greater than'").

2. *Make explicit that orientation is an important factor in distinguishing written forms.* Experience has taught children that orientation is not crucial for identifying most objects. Mommy is Mommy whether you look at her from the front or back, whether your view is right side up or upside down. Young children then may overlook orientation and not distinguish between written forms such as the numerals 6 and 9, the operational signs $+$ and $\times$, or letters like *b* and *d*. Add to this the difficulty that some children have in mastering directional vocabulary to label these orienta-

tions, and it is not surprising that reversal errors are very common at the primary level.

3. *Writing training should focus on pointing out and mastering a motor plan.* It may help to note explicitly where to start, what direction to go in, and where to stop and change direction. To make the exercise instructive and useful, tracing should be accompanied by instruction on motor plans.

4. *Encourage self-rehearsal of the motor plans.* Especially for verbally inclined children, learning may be facilitated by encouraging the pupil to describe the distinguishing characteristics and the motor plan themselves.

A Demonstration Project with Mentally Handicapped Children

A cognitive approach to writing numerals can be used even with mentally handicapped children. A demonstration study was undertaken with five moderately and three mildly mentally handicapped pupils. The training used special writing paper from the Recipe for Reading Program (Traub, 1977) that has markings to assist in giving direction (see Figure 11.2). To assist in giving right-and-left directions, a drum and stick are placed in the upper-left-hand corner and a ball and bat in the upper-right-hand corner. To assist in giving up-and-down direction, each row on the paper begins with a small house. The top line of the row is drawn from the top of the roof, the middle line extends from the junction of the roof and house, and

FIGURE 11.2: Writing Paper from Recipe for Reading*

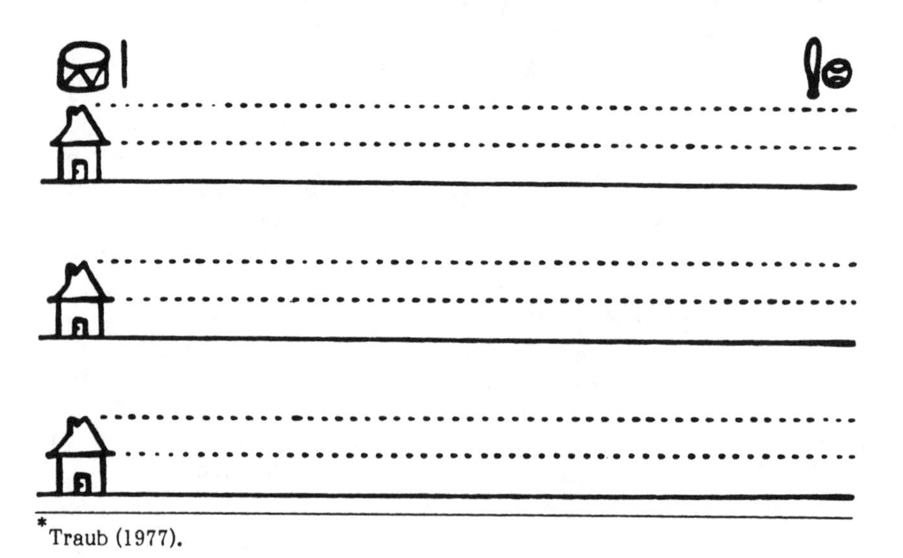

*Traub (1977).

the bottom line runs from the bottom of the house (along the ground). The lines are named "top," "middle," and "bottom" (or "ground"). Children first learned the component parts of the paper so that they could immediately recognize where to start and stop when given instructions.

Instruction then focused on providing the children with an explicit motor plan for the numerals 1 to 10 (see Example 11.2). The instructor slowly drew a numeral while describing the motor plan. After describing and demonstrating the motor plan, the instructor talked the child through the process. The instructor would define the starting point for the child by drawing a dot. Then the child would practice writing the numeral as the instructor described the motor plan. If the child experienced difficulty the instructor would take back the pencil, place the child's hand over his or her own, and proceed through the motor plan while describing it. This hand-over-hand technique was used until the child could write the number independently. The child was asked to describe the essence of the motor plan before, during, and after practice efforts. (The children were not required to repeat the motor plan verbatim.)

The training proved to be quite successful in improving numeral-writing performance. For example, Steven, the boy described in Example 11.1, eagerly and quickly learned the motor plans. Indeed, to and from class, Steven would trace numerals in the air, as he described their execution. Verbalizations of his motor plans were quite frequent with new, unfamiliar numerals but decreased as he mastered the numerals. Steven's trainer and teacher were concerned that there might be a problem discontinuing the special writing paper used in his training. However, after the training, Steven and the other children performed well on blank, unlined paper. Apparently the children had developed a mental motor plan that was not tied to the cues for position and direction provided by the special training paper.

SUMMARY

School requires children to recognize, read, and write formal symbols. Learning difficulties typically do not stem from inadequate practice or perceptual-motor difficulties. Problems in recognizing or reading numerals can more often be traced to not knowing the defining characteristics of the numerals. Thus remedial efforts should focus on making explicit the parts and part-whole relationships that distinguish, say, each numeral. Writing problems typically can be traced to a deficient or faulty motor plan. Therefore, remediation should center on making explicit the rules for making numerals or other symbols.

EXAMPLE 11.2

A Cognitive Approach
to Numeral-Writing Instruction

Because instruction was designed to build on previously learned numerals, the numerals were presented in the following order. Easily confused numerals, such as 5 and 2 or 6 and 9, were taught one after the other, and differences in their execution were highlighted.

1. "One starts at the top line and drops straight down to the bottom line (ground). It goes all the way to the ground but not below it."

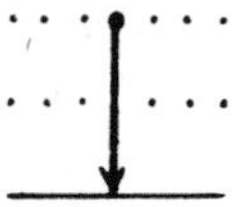

7. "Seven starts at the top line, walks along the top line toward the bat-and-ball side, stops, and then slides down toward the bottom line and back toward the drum-and-stick side. Like the one, it goes all the way to the ground but not below it." Note that some children may lift the pencil between the two strokes. This produces results like -/ . Therefore, emphasize keeping the pencil in contact with the paper.

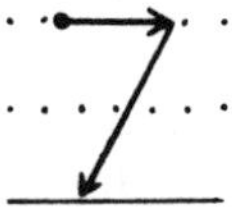

4. (1) "Four starts at the top line, drops straight down to the middle line, stops, then walks along the middle line toward the bat-and-ball side, and stops. (2) Then another line starts at the top line and drops straight down to the ground, crossing the first line that goes along the middle." Note that it is important to emphasize the stopping points for some children.

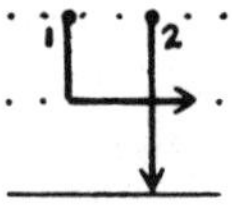

5. (1) "Five starts at the top line, drops straight down to the middle line, stops (like the four), and then makes a big tummy toward the bat-and-ball side that ends up on the ground. (2) Five also has a hat. Go back to the top line where you started, walk along the top line toward the bat-and-ball side, and stop."

2. "Two starts below the top line (at the chimney); now it makes an ear by traveling up toward the ball-and-bat side, touching the top line, curving around all the way down to the bottom line, stopping underneath where it begins; and then it adds a tail by walking along the bottom line toward the ball-and-bat side."

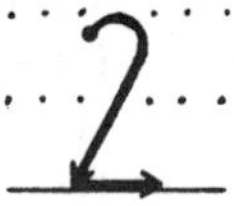

EXAMPLE 11.2 (continued)

3. 

"Three starts below the top line (at the chimney like the number two) and travels up toward the ball-and-bat side, touches the top line, makes a tummy that rests on the middle line, makes a bigger tummy toward the ball-and-bat side that touches the bottom line (ground), and curls up toward the drum-and-stick side."

10. "Ten is two numbers together. (1) First draw a one. (2) Now pick up the pencil and move it toward the ball-and-bat side. Next draw a circle that starts at the top line and travels toward the drum-and-stick side all the way down to the ground and back up toward the bat-and-ball side to where it began." If one motion for the zero is too difficult, have the child draw two connecting half circles.

6. "Six starts at the top line, curves down toward the drum-and-stick side until reaching the middle line, continues to curve down but now toward the bat-and-ball side until it reaches the bottom line, and then makes a ball at the bottom that touches the middle line before it closes." If a curve is too difficult, try a stick and ball. Specifically: "Six starts at the top line, drops straight down to the ground (just like the one), and then adds a ball that comes up toward the bat-and-ball side, touches the middle line, and closes."

9. "Nine starts below the top line (at the chimney), makes a ball above the middle line that first goes toward the drum-and-stick side and returns to the start, and then curves down toward the bat-and-ball side until it reaches the bottom line (ground). So six has a ball at the bottom that faces the ball-and-bat side, and nine has a ball at the top that faces the drum-and-stick side." If a curve is too difficult, use a ball and stick approach. Specifically: "Nine starts below the top line (at the chimney), makes a ball above the middle line that first comes toward the drum-and-stick side and then returns to the start. Now, from where you started, a line drops straight down to the "bottom line (ground)."

8. 

"Eight starts at the top line and it makes a ball that touches the middle line and then closes. Then it makes a second ball (circle) that touches the first one at the middle line, travels all the way down to the ground, and finishes by closing. (It looks like a snowman)." Note that some children make the bottom circle considerably smaller than the top. The snowman image sometimes helps.

CHAPTER 12

Place-Value Numeration Skills and Concepts

Why do children have difficulty reading and writing progressively larger numerals? What kinds of errors do children make in reading and writing multidigit numerals, and why? What are the key place-value concepts that children must master to understand our numeration system? How can teachers build on children's existing knowledge to foster a secure learning of multidigit numeration skills and concepts?

READING AND WRITING MULTIDIGIT NUMERALS

Adam's Place-Value Numeration Difficulties

When the case study began, 11-year-old Adam did not appear to have any understanding of place value (Baroody, 1983a). Indeed, this learning-disabled boy could not even correctly label the ones and tens place of a two-digit numeral. He could write two-digit numerals but still had great difficulty with larger numerals. For example, in our eighteenth session (9/2/81), Adam wrote 1013 for "one hundred thirteen," 1017 for "one hundred seventeen," and 2002 for "two hundred two." Four months later (Session 22, 1/13/82), Adam still had not learned to write three-digit numerals (e.g., he wrote 1003 for "one hundred three").

Like Adam, do most children begin to read and write multidigit numerals before they understand place value—the underlying rationale for these procedures? Why was there such a clear breakdown in his performance at the three-digit level? Do children typically master reading and writing two-digit numerals but not three-digit numerals, three-digit numerals but not four-digit numerals, and so forth? Why did Adam regularly include an extra zero in writing three-digit numerals? Can special education children like Adam learn place-value skills and concepts?

Analyses of the Learning Tasks

In a very real sense, multidigit numerals are a number *sentence*. A sentence is not merely a string of words, each with its own meaning. Rather, it encodes relationships among words. Compare the following sentences:

Jim hit the ball.
The ball hit Jim.

Though both sentences contain the same words, the relationships among the words are different. A sentence must be read as a whole to learn its meaning—the particular relationships it conveys. Likewise, a multidigit numeral is not simply a string of individual single-digit numerals, each with its own meaning. A multidigit numeral is a number sentence that encodes relationships among individual digits to signify *a* number (McCloskey, Caramazza, & Basili, 1984). Thus multidigit numerals must be read as a whole to learn their meaning (e.g., 47 is not read as "four seven" but as "forty-seven"). Though adults may take for granted that multidigit numerals encode relationships, this may not be apparent to a young child.

The relationships among individual digits are encoded by the position of the digits. For instance, the order of the digits in 47 specifies a different meaning than the order 74. Moreover, the place of the four in 47 and 470 conveys different meanings ("forty" versus "four hundred"). Again, though it is obvious to adults that order and place encode information, it is by no means apparent to young children.

To read and write multidigit numerals, children must recognize that these symbols are really number sentences. Moreover, they must learn how relationships are encoded and decoded. This entails learning how our highly regular written system maps onto (does and does not correspond to) our verbal number sequence. In effect, children have to learn procedures for translating one system into another—including special rules for exceptional cases.

READING. Like reading a word sentence, reading a multidigit numeral is a decoding process. To read a numeral, a child must decode the information specified by position (order and place). The first step is noting the number of digits present (McCloskey et al., 1984). For example, an initial assessment of 47 indicates that the numeral has two digits. The second step entails specifying the relationships among the digits. Two digits specify that the first (left-hand) slot is filled by a term drawn from the "tens group" (the decade sequence) and the second (right-hand) slot by a term from the "ones group" (the one-to-nine number sequence). This step in reading two-digit numerals is summarized below:

______	______
TENS	ONES

The third step consists of filling in each slot with the specific and appropriate number name. In the case of 47, the tens slot is filled by recalling the verbal label "forty" from the decade sequence, and the ones slot is filled by recalling "seven" from the one-to-nine number sequence:

forty	seven
TENS	ONES

This procedure for translating a written symbol into a verbal form applies to reading all two-digit numerals except for the teens (numerals with a 1 in the tens place) and decades (numerals with a zero as a placeholder). With a teen term such as 17, the tens digit simply serves to mark the numeral as a "teen." A reader silently notes its presence, focuses attention on the ones digit (7), and calls up the corresponding verbal representation indicated by the ones-place digit ("*seven* + teen"). To decode 11, 12, 13, or 15, the child must recognize the terms as "irregular" teens and recall their special (rotely learned) names: "eleven," "twelve," "thir + teen," or "fif + teen." With decades, the 0 in the ones position simply serves to mark the numeral as a decade. With 40, say, a reader silently notes the presence of the 0, focuses on the tens term 4, and calls up the corresponding decade term: "For + ty."

With three-digit numerals, the reading procedure is more complicated. The initial assessment must correctly identify the presence of three digits. Because the left-most digit specifies the number of hundreds, a child must learn an additional rule for decoding this relationship. Specifically, a child must know that this first digit is drawn from the ones group and joined by the place-designation name "hundred." This process is depicted below:

______	+ hundred	______	______
ONES		TENS	ONES

To read, say 247, the number name "two" is called up and supplemented with the "hundred," "forty" is called up to fill the middle slot, and "seven" the right-hand slot:

two	+ hundred	forty	seven
ONES		TENS	ONES

Again the exceptions to this straightforward procedure involve teens and zeros. The teen and decade exceptions continue to apply. For example, 217 is read: "two + hundred, [note and postpone translating 1 as 'teen'], seven + teen," and 240 is read: "two + hundred, [note but do not mention the marker 0], forty." In addition, zero can also serve as a placeholder in the tens place. This necessitates an embedded-zero rule: "Embedded zeros are not mentioned but specify that the slot is skipped" (e.g., 207: "two hundred seven").

Learning to read larger numerals involves learning additional rules for decoding these more complex relationships. In particular, reading four-digit numerals entails learning the place-designation name "thousand," when this number name is applicable, and how to use it (e.g., 3,247: "three + thousand, two + hundred, forty, seven"):

three	+ thousand	two	+ hundred	forty	seven
ONES		ONES		TENS	ONES

For four-digit numerals, children also must take care in using an embedded-zero rule. That is, they must be careful about what place-value designation to start up with and which to pick up with again after the zero. For instance, for 1,070, a child needs to start with the place-value designation "thousand" not "hundred" and finish with a tens designation ("seven*ty*") as opposed to a hundreds designation ("seven *hundred*") or a ones term ("seven").

Reading five- and six-digit numerals does not require learning an additional place-designation name, but it does require recognizing the applicability of previous procedures and rules. In the case of 583,247, the first three digits are read as though it were just any three-digit numeral (except that the place-designation name "thousand" is added). The same is true with even larger numerals (e.g., the italicized portions in *247*,590,130 and *247*,683,590,130 are also read as "two hundred forty-seven"). Thus, once children have mastered the procedures and rules for reading numerals up to three digits, they only have to learn the appropriate place-value designations to read larger and larger terms: one comma (four to six digits) signifies "thousands," two commas (seven to nine digits) signifies "millions," and so forth.

WRITING. Learning to write numerals requires encoding a verbal number into appropriate written symbols. For a two-digit number, like "forty-seven," the tens term should specify two and only two digits: 47, not 407. The place-designation term "hundred" specifies three and only three digits (e.g., "two hundred three" is represented by a 2 followed by two more digits, not 2003 or 23).

Typical Progress and Common Pitfalls

This cognitive analysis helps to explain the typical developmental pattern of these skills and the errors common to young children or children having learning difficulties.

SKILL BEFORE UNDERSTANDING. Like Adam, described at the beginning of the chapter, children typically begin to read and write multidigit numerals before they have much, or even any, understanding of place value (Ginsburg, 1982). If children do not know the place-value designations or the rules for decoding/encoding the more complex relationships involved, they will be unable to read/write larger multidigit numerals. For example, even when children learn how to read and write two-digit numerals, they will not be able to read three-digit numerals until they have learned the place-value designation "hundred" and the place-value based embedded-zero rule. Because he did not realize that the number of digits encodes value (three digits specifies hundred, four digits specifies thousand, etc.), Adam had great difficulty writing numerals beyond the two-digit level.

TYPICAL LEVELS OF MASTERY. As with many other mathematical skills, then, children learn to read and write numerals by degrees (Baroody, Gannon, Berent, & Ginsburg, 1984). By the end of kindergarten, most children can read teen numerals. However, reading two-digit numerals necessitates learning a new decoding rule for mapping the written symbols to the verbal sequence. Moreover, it demands recalling decade information, which in many cases has not yet been learned. Many kindergartners have not solved the decade problem and thus cannot count by tens. Others may not have sufficiently internalized this knowledge to use it in new tasks.

During first grade, children master the number sequence to 100 and the rules for decoding two-digit numerals. Reading larger numerals depends on and thus must await sufficient counting experience or reflection to construct the rules for generating the hundreds, the thousands, and so forth. The amount of direct counting experience, and hence the degree of familiarity, drops off dramatically as numbers increase in size beyond 100. For example, a child might have an informal notion that a million is a large number, but more exact knowledge may depend on formal instruction: A million is more than a thousand but less than a billion (relative value); millions consist of seven, eight, or nine digits (relative size); and the second comma from the right in a term, like 7,222,333, designates the name "million" (place-value designation). Thus by the end of first grade, children typically learn to read two-digit but not three-digit numerals. Second graders usually master three-digit numerals by the end of the year but not larger terms. And four-digit numerals typically are not learned until third grade.

Children also typically master numeral writing according to size level. By the end of kindergarten, about half the children in a typical class are able to write the teens or other two-digit numerals. Children may master writing two-digit numerals before teen numerals. Mastery of the teens may be delayed because of difficulty in learning the irregular number-sequence terms "*thir*teen" and "*fif*teen" or because the procedural rules for encoding (writing) the teens are different from other two-digit numerals. First graders usually master writing the teens and other two-digit numerals but not three-digit numerals. Second graders commonly master three-digit numerals, but only about half also learn to write four-digit numerals.

COMMON ERRORS. An analysis of the decoding and encoding processes explains why children's numeral-reading and -writing errors are quite often systematic. Incomplete or incorrect decoding/encoding rules can lead to various types of errors. These deficiencies frequently arise because of the lack of place-value knowledge. Children who do not realize that multidigit numerals are a number sentence that must be treated as a whole may decode and read the numerals separately (e.g., read 27 as "two seven"). This is a common reading error among kindergartners (Baroody et al., 1984). Children who do not understand that position encodes information may disregard digit order (e.g., sometimes read 71 as "seventeen") or overlook the number of digits (e.g., read 1,047 as "one hundred forty-seven").

Likewise, incomplete or incorrect decoding rules can lead to various errors in writing multidigit numerals. Young children are especially likely to make errors with the teens and decades. Because the decoding rule for writing teens conflicts with that of writing other multidigit numerals, reversing the digits of a teen (e.g., writing "nineteen" in the order in which the digits are heard: as 91) is a common error among kindergarten children. Because decade terms are written with a 0 (e.g., "forty" is written 40), some children may assume that a 0 should be written whenever they hear a decade term (e.g., write "forty-two" as 402).

Children may have particular difficulty reading and writing numerals with zeros, because they are not prepared to view zero as a placeholder. Children are accustomed to treating multidigit numerals, such as 10, 70, 100, 300, as undifferentiated wholes: ten units, not 1 ten and 0 ones; seventy units, not 7 tens and 0 ones, and so forth. (Baroody et al., 1984).

Thus, when they see a multidigit numeral such as 4002, they naturally assume that the 4, 0, and 0 are linked and stand for four hundred units. This could also account for why some children read multidigit numerals as they are written (e.g., reading 402 as "forty-two" or reading 4,002 as "four hundred two").

It also explains why many children write numerals with unneeded zeros (e.g., writing "four hundred two" as 4002). For example, many first graders run into difficulty with three-digit numerals because they mechanically write 100 whenever they hear "one hundred" (e.g., writing "one hundred forty" as 10040 or "one hundred two" as 1002). Many second graders make similar errors with four-digit numerals.

Children may have difficulty treating zero as a placeholder for another reason. Children are accustomed to interpreting zero as meaning "nothing" (as referring to an empty set). This interpretation is in direct conflict with the decoding and encoding rules for embedded zero. For example, in attempting to read 4,002, a child may interpret the zeroes as "nothing to notice," ignore them, and read the numeral as "forty-two" (Kamii, 1981). Similarly, in an effort to write "four thousand two," a child may simply write 42.

Inadequate number-sequence and place-designation information can also lead to systematic errors. If children do not know the appropriate number name, they typically decode the novel numeral in terms of what they do know. For example, even kindergartners who can read teen numerals may be unfamiliar with the rule-governed portion of the number sequence. As a result, they may treat an unknown two-digit numeral as two known numerals (e.g., "two, seven" for 27), substitute a known for an unknown term (e.g., "seventeen" for 47), or make up a name based on known numerals (e.g., "twenty-seventeen" for 2,073). First or second graders often substitute a smaller familiar term for an unfamiliar one (e.g., "four hundred seventy-three" for 4,073) or read unfamiliar terms as a combination of familiar terms (e.g., "one hundred two" for 1,002).

PLACE-VALUE KNOWLEDGE

Development

Knowledge of our base-ten, place-value numeration system develops gradually and builds upon previous counting knowledge (Baroody et al., 1984; Resnick, 1983). Children learn relatively early and quickly to recognize the ones, tens, hundreds, and even thousands places—that is, the names and locations of the ones place, tens place, and so forth (e.g., for

425, the 5 is in the ones place, the 2 is in the tens place, and the 4 is in the hundreds place). Children also quickly master place-value designations—that is, they can identify how many ones, tens, hundreds, and so forth a multidigit numeral represents (e.g., 425 stands for 5 ones, 2 tens, and 4 hundreds). However, such skills are frequently learned in a rote fashion and represent a rather superficial understanding of our base-ten numeration system (Resnick, 1982).

An important place-value concept for children to grasp is that zero acts as a placeholder. In other words, children have to go beyond thinking of zero as meaning "nothing" (as in an empty set) or having no effect (as in $7 + 0 = 7$), and think of it as indicating an "empty column" (e.g., in 402, the zero represents no groups of ten). In comparison to children's initial meanings for zero, this place-value meaning of zero is rather abstract.

Deep knowledge of our place-value system involves understanding the structure of the base-ten numeration system (Baroody et al., 1984). A key to this understanding is that there are base-ten equivalents: 10 ones can be exchanged for 1 ten, 10 tens for 1 hundred, 10 hundreds for 1 thousand, and so forth. Another key to understanding the structure of the system is that there is a repeating pattern with predictable transition points. That is, our base-ten system is made up of 10 single-digit symbols (0, 1, 2, 3, 4, 5, 6, 7, 8, and 9) that are systematically combined to form two-digit numerals beginning with 10 and ending with 99. In a similar manner, the three-digit numerals repeat the pattern beginning with 100 and ending with 999, as do the four-digit numerals (1,000 to 9,999), and so forth. Recognizing this pattern—stating the smallest and largest one-, two-, or three-digit numeral—is an advanced skill.

Consider the following exchange with Adam during Session 26 (2/24/82).

INTERVIEWER: What's the smallest one-digit numeral in our number system?
ADAM: Zero.
INTERVIEWER: What is the largest one-digit number?
ADAM: Ten.
INTERVIEWER: One digit.
ADAM: One hundred.
INTERVIEWER: How many digits are in 100?
ADAM: Three.

After several more guesses including 11, 12, and 21, the interviewer guided Adam to the correct answer: Nine is the largest one-digit numeral. Asked then what the smallest two-digit numeral was, he responded correctly: "Ten." However, he again needed help to conclude that 99 was the largest two-digit numeral. It should be noted that this was not the first time Adam had been asked such questions. He had been quizzed and given instruction on this topic over a period of three weeks (Sessions 23, 24, and 25). Understanding how our base-ten, place-value numeration system is organized can be a very difficult step. Indeed, one study (Baroody et al., 1984) found that only 21% of the second graders and 76% of the third graders tested could correctly identify the smallest and largest one-, two-, and three-digit numbers. (Most children faltered at the three-digit level—erroneously giving 199 or 900 as the largest three-digit number. Note, though, that such an error indicates a sensible intuition about the three-to-four-digit transition point.)

With development, children can think about multidigit numbers with more flexibility. An understanding of multiple partitions involves, for example, recognizing that 43 may be partitioned into 3 tens and 13 ones as well as 4 tens and 3 ones. Resnick (1983) hypothesizes that this is a prerequisite concept for understanding carrying and borrowing procedures.

Applications

Thinking in terms of units other than ones gives children flexibility in dealing with a range of tasks (Payne & Rathmell, 1975). Consider complex enumeration: counting with different units. Children frequently first encounter complex enumeration in the context of combining ten(s) and ones. For example, the problem of adding two dimes and three pennies could be solved by first counting by tens and then by ones: 10, 20, 21, 22, 23. Switching between two counts or among three or more counts is frequently quite difficult for primary-age children. For example, though third graders can count, say, four $100 bills by hundreds, count three $10 bills by tens, or count two $1 bills by ones, many have difficulty switching from one count to another when the different denominations are combined and all three counts are required (Resnick, 1983). Some children persist with only one type of count. For instance, they correctly count the four $100 bills ("100, 200, 300, 400") but then continue to count by hundreds when they count the three $10 bills and two $1 bills ("500, 600, 700, 800, 900"). Others start with a hundreds count and then incorrectly switch to ones. For example, they correctly count the four $100 bills by hundreds ("100, 200, 300, 400") but count the three $10 bills as well as the $1 bills by ones ("401, 402, 403, 404, 405"). Some children overlook the differences in denominations and simply count all the bills by ones: "1, 2, 3, 4, 5, 6, 7, 8, 9."

EDUCATIONAL IMPLICATIONS: MEANINGFUL PLACE-VALUE NUMERATION TRAINING

1. *Consider the effects of numeral size.* The increasing complexity of the decoding/encoding rules and the number-sequence data required to read and write progressively larger numerals tends to produce level-by-level mastery. Thus number size clearly needs to be taken into account when planning numeral-reading or numeral-writing instruction or testing. A child who is having difficulty with school mathematics involving three-digit numerals may have unsuspected strengths when asked to work with two-digit numerals. On the other hand, it cannot be taken for granted that skills successfully learned with one- or two-digit numerals will necessarily carry over to three- or four-digit numerals.

2. *Examine errors to guide remedial efforts.* Many numeral-reading and -writing errors are not random or senseless but the result of active attempts to use existing knowledge to make sense of or cope with unfamiliar numerals. Because numeral-reading and -writing errors are often systematic, they can provide educators with an important vehicle for understanding a child's incomplete or incorrect procedural rules or place-designation information.

3. *Explicitly point out decoding/encoding rules.* One method of making explicit the importance of order in defining numerals is to have a child compare numerals with oppositely ordered digits. For example, with a child who writes "fifteen" as 51, point out that "fifty-one" is written 51 and that we have to write fifteen and fifty-one in different ways so as not to confuse them.

Make explicit the place-keeping role of zeros. For the decades, such as 40, children can be taught that the 4 specifies "forty" and the 0 means, "No need to read further; no number is needed from the ones group." Note that such a procedure is more consistent with how other two-digit numerals are read. For larger numerals, point out the embedded-zero rule—that a zero does not mean "nothing to notice."

Explicit instruction becomes increasingly important as numerals become larger. One learning aid for reading numerals of four digits or more is use of the comma (or commas) to partition the whole into more manageable chunks. Once a child has isolated a chunk, the digits can be read as the more familiar one- to three-digit numerals. For example, the portion to the left of the comma in *1*,003 is simply read "one"—plus the place-name designation "thousand." With the numeral 23,725, the chunk to the left of the comma is simply read as "twenty-three" (plus the place-value designation "thousand").

4. *Use an informal approach to cultivate place-value numeration skills and concepts together.* Learning problems—especially long-lasting problems—may well be due to inadequate knowledge of place value. The following subsection describes case-study work (Baroody, 1986) with learning-disabled, mentally handicapped, and underachieving children. These case studies highlight the value of using an informal approach to teach numeration skills and place-value understanding simultaneously. Combining skill and meaningful instruction should reduce considerably numeral-reading and numeral-writing learning difficulties and errors. This may be especially important for special education children such as Adam. Moreover, the case studies describe games and activities that not only are helpful in promoting learning but can be a great deal of fun.

An Informal Approach to Remediation

COMBINING INSTRUCTION ON PLACE-VALUE SKILLS AND CONCEPTS. Remedial efforts with Adam focused on teaching basic place-value skills and concepts simultaneously. Initial efforts were directed at (a) identifying the place names and values of numerals up to three digits; (b) thinking of 10 elements as *a* group of ten and 10 tens as *a* group of hundred (recognizing base-ten equivalents); (c) viewing double-digit numerals as composites of tens and ones; and (d) connecting concrete representations of base-ten and written numerals including the use of zero as a placeholder.

Forward Bowling provided an entertaining basis for attacking all four objectives delineated above. This activity involved keeping track of the score of a "skittleball" match (a miniture bowling game). To keep the scoring straightforward, strikes and spares are only scored as 10 points (Wynroth, 1969/1980). Three scoring variations were used. In the first (concrete) method, Adam would (a) count up the pins that had been knocked down that frame; (b) make an equivalent number of marks above the scoring box for that frame; (c) check to see if there were 10 marks and circle them if there were (uncircled marks from earlier frames could be used); (d) count all circled marks by tens and any remaining marks by one to determine the score for the frame; and (e) enter the numerical score in the scoring box for that frame (see Figure 12.1, Frame A).

A second concrete version involved using sets of 10 interlocking blocks of different colors. Dienes blocks are an excellent alternative (see, for example, Resnick & Ford, 1981). Each participant had a scoreboard consisting of a green dish labeled "ones" and, to the left of that, a pink dish labeled "tens." (a) After each frame, the number of fallen pins was tallied. (b) An equivalent number of blocks were counted out and placed in the "ones" dish. (c) The blocks in the "ones" dish were then counted; if the total

FIGURE 12.1: Scoring Procedures for Bowling

A. Slashes to Numerals (Concrete Technique)

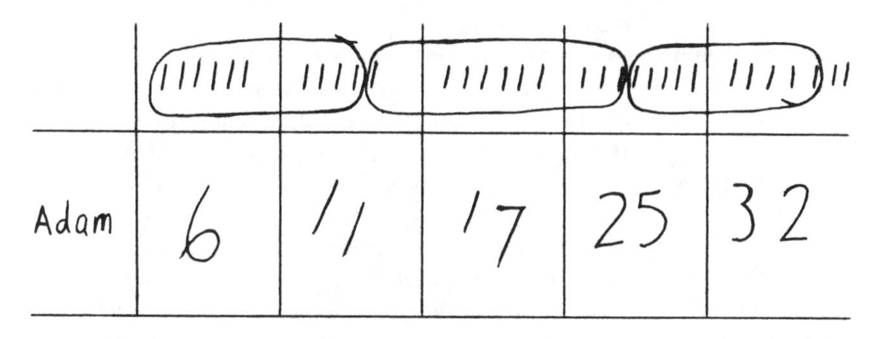

B. Blocks to Numerals (Concrete Technique)

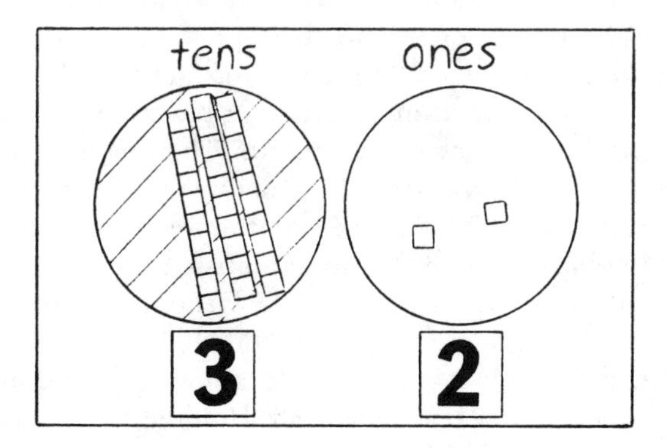

C. Pegs to Numerals (Semiabstract Technique)

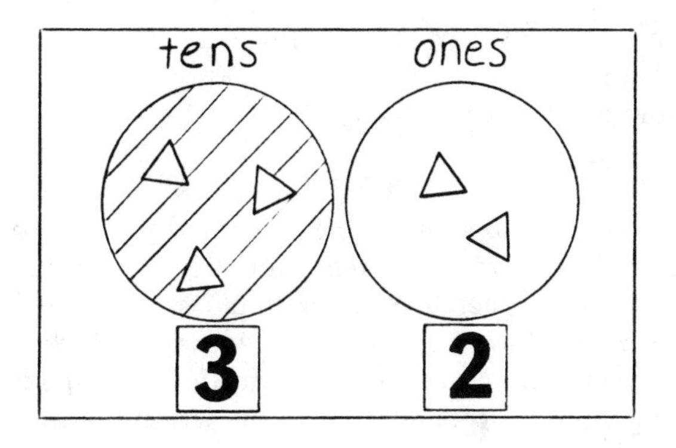

exceeded nine, 10 blocks were put together and placed in the tens dish as "*a* ten-bar." (d) Adam then placed a numeral card under each dish to correspond to the number of tens and ones and read the score (see Figure 12.1, Frame B).

The third and most abstract method comes from the Wynroth (1969/1980) curriculum and uses pegs to represent both ones and tens. The same procedure as Method 2 was used except for Step 3. At each turn, the ones dish was checked to see if there were enough to make a ten. If so, the 10 pegs were removed, and a single peg was then put in the tens dish. The interviewer would comment during this process, "We'll trade these 10 ones for 1 ten." (See Figure 12.1, Frame C.) In the second and third scoring systems, the numeral zero is used as a placeholder—to indicate no ones. In more advanced games, a "hundreds dish" can be introduced. (Many other devices can be used for scoring, such as the variation that will be seen in Figure 13.3.)

The following incident transpired more than 4 months after Adam had mastered the first two scoring techniques and without further practice in the interval. With a score of 13 already (represented by three interlocking blocks in the "ones dish" and a ten-bar in the "tens dish"), Adam knocked down eight more pins. He counted out eight blocks and added two blocks from his ones dish while saying, "8, 9, 10." Then he announced, "Now I have two tens: twenty." (Note that he spontaneously equated the base-ten representation of "two tens" with the count representation of "twenty.") He put the new ten-bar in the tens dish and labeled the dish with a numeral "2" and the "ones dish" with a numeral "1." Hence he readily connected the concrete representation with its written symbol. Then later, with a score of 26, he knocked down nine more pins. He obtained the appropriate number of blocks and made a new (third) ten-bar. Holding the third ten-bar in his hand, he announced, "Thirty . . . [counted the remaining ones blocks] five." On his next turn, he added 7 more to his total and announced: "42."

INTERVIEWER: You have 42 there? How many tens?
ADAM: Four.
INTERVIEWER: How many ones?
ADAM: Two.

Adam, moreover, could operate in the opposite direction: Given a numeral, he could indicate how many tens and ones it represented as well as make the appropriate concrete model with blocks.

The Target Game has as its objective viewing multidigit numerals as composites of ones, tens, hundreds, and so forth; mastering base-ten equivalents; and providing a concrete model for writing numerals, including zero as a placeholder. One version of the game involves throwing beanbags at a target. The target, which can be outlined in masking tape on the floor, consists of three concentric circles labeled 1, 10, and 100. (A fourth circle labeled 1,000 can be added for advanced players). Figure 12.2 illustrates the scoring procedure. (a) Of six beanbags thrown on this turn, two landed on the 100 bull's-eye and three landed on the field marked 10. (b) The child puts 2 pegs in the hundreds dish and three in the tens dish. (c) Because there were already seven pegs in the tens dish, the child can exchange 10 in this dish for one in the hundreds dish. (d) The child then labels his score with a number. A second version, somewhat messier but very exciting, involves dropping pennies onto a target submerged in water. The target can be made up of plates of various sizes.

Trading Up focuses specifically on the objective of viewing double-digit numerals as composites of tens and ones and learning the base-ten equivalents. Each player collects money and whenever possible trades it in for a higher denomination bill. The first player to get a $100 bill in trade is the winner. On each turn a player rolls dice to see how much money to collect. To keep the focus on base-ten equivalents, only $1, $10, and $100 bills are used. Rolls of 10, 11, and 12 provide an opportunity for a player to immediately collect a higher-denomination ($10) bill and to see that 10 one-dollar bills are equivalent to 1 ten-dollar bill and that teens (such as 12) are composites of a ten and (two) ones. Every time a player collects 10 one-dollar bills they can be traded.

A more advanced trading-up game was used with Adam. In this version, $1, $10, $100, and $1,000 bills were used. The winner was the first to obtain a $1,000 bill. On each turn, a player rolled numeral dice to see how much money he could collect. The player could arrange the dice to make the larger number. For example, if 2 and 6 were rolled, a player could take $26 or switch the order of the dice and take $62.

Hieroglyphics training can also provide interesting learning and practice exercises for many place-value objectives. The Egyptian hieroglyphics are a straightforward base-ten system. Adam was first exposed to the Egyptian symbols for 1 (|), 10 (∩), and 100 (𓍢). (See Bunt et al., 1976, for a more complete description.) Adam was then asked to write "2 through 9." Because Adam was unsure how to proceed, I demonstrated the hieroglyphic for 2 (||). Adam responded, "Oh, I see," and filled in the hieroglyphic equivalents of 3 to 9. For 10, he laboriously made 10 marks. I pointed out that there was an easier way of writing 10 (∩). Asked to write the Egyptian equivalents of 10, 20, 30, . . . , 90, he proceeded without difficulty. Asked to translate 12, Adam was unsure what to do. He was helped to induce the correct form (∩||). He then quickly translated 13 (∩|||), 23 (∩∩|||), and 43 (∩∩∩∩|||). Asked to translate 100 and 103, he overlooked the

FIGURE 12.2: Target-Game Scoring

A. Situation just prior to scoring

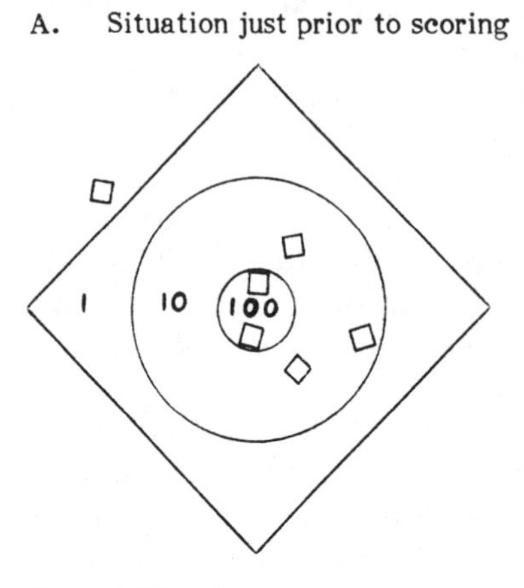

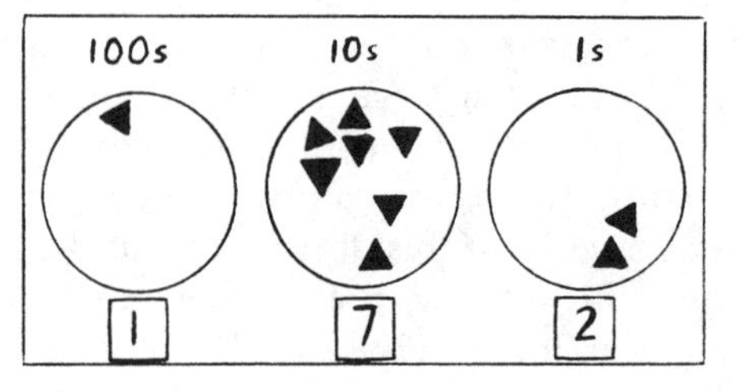

B. Adding the new scores

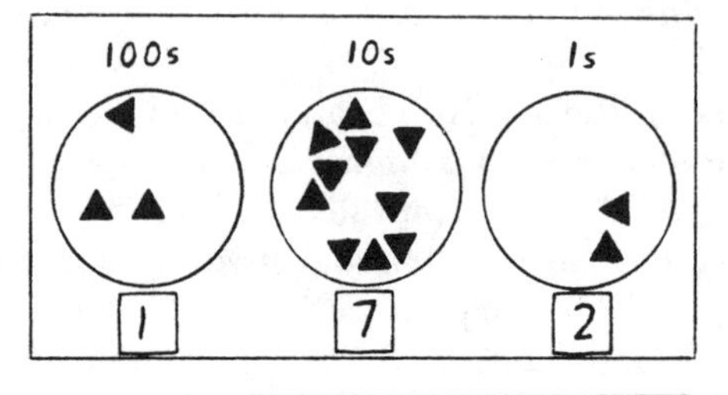

C. Exchanging

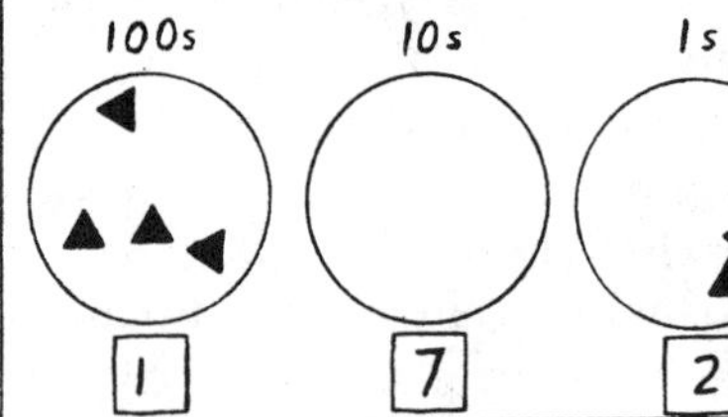

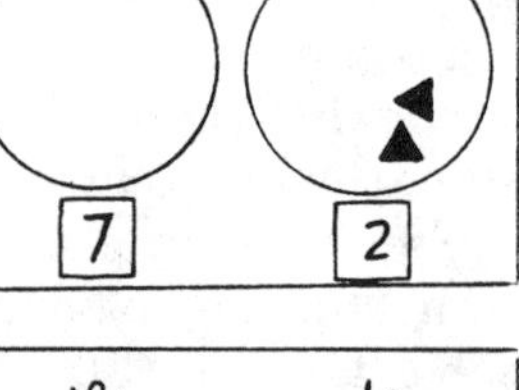

D. Relabeling

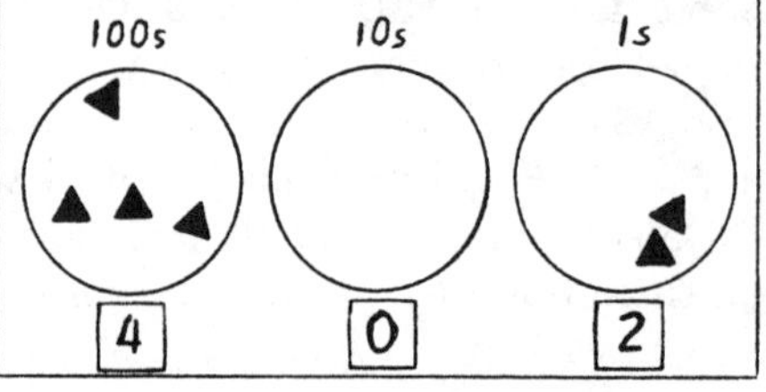

symbol for 100 (ᕦ), so I pointed out this abbreviation to him as well. Later Adam efficiently interpreted such relatively difficult hieroglyphics as ᕦI (101) and ᕦ∩I (111) as well as ∩I (11) and ∩∩∩∩∩II (52).

INTRODUCING OTHER BASE SYSTEMS. One reason children are introduced to other base systems is so that they develop a better understanding of the base-ten, place-value numeration system. Unfortunately, such instruction is often introduced in a highly formal manner that children cannot comprehend. As a result, material that was intended to promote meaningful learning is, if learned at all, rote memorized and added to the storehouse of uncomprehended mathematical facts.

Introducing other base systems can be done in an informal and meaningful way. Because children become familiar with our everyday (base-ten) system through counting, it might be helpful to use counting to introduce new base systems. If a teacher was interested in introducing base two, for example, he or she could ask: "Our (base-ten) number system has how many building blocks—different numerals that can stand by themselves to represent a number or be combined to represent other numbers?" In base ten, there are 10: 0, 1, 2, 3, 4, 5, 6, 7, 8, 9. Other numbers can be represented by combining these basic building blocks. Now the teacher can pose the question: "What if we had a number system which had only two building blocks? Design such a number system and show how you would count to 20 in it." The count sequences can be generated by means of a few simple principles, which the children may need some hints to discover or reminders to follow. Rule 1: Count as far as you can with the standard string of building blocks. In base two, this will not take you very far. The basic blocks (0, 1) are exhausted representing zero and one. Rule 2: Then combine the building blocks in the *simplest* way possible. In base two, two must be represented by a two-symbol combination, and the simplest combination of building blocks is 10 (one and zero cannot be combined in the reverse order because 01 is equivalent to 1). Now simply continue to apply Rule 2. Accordingly, three is represented by 11. Four must be represented by three digits because all two-digit combinations are used; the simplest three-digit combination is 100. (Note 10 or 100 is not read as "ten" or "one hundred" in base two, but may be referred to as "one-zero" and "one-zero-zero," respectively.) Various answers might be correct depending on a student's choice of how to represent zero and one. (For example, * and # or numerals other than 0 and 1 could be employed.) Some students may even hit upon the standard base-two representations (see column 2 of Table 12.1). Without introducing formal terms such as "base two," this approach was used with capable third graders and appeared to be readily understood.

TABLE 12.1: Base-Ten, Base-Two, and Base-Eight Representations

Quantity	Base-Ten Symbol	Base-Two Symbol	Base-Eight Symbol
	0	0	0
.	1	1	1
..	2	10	2
...	3	11	3
....	4	100	4
.....	5	101	5
..... .	6	110	6
....	7	111	7
....	8	1000	10
....	9	1001	11
....	10	1010	12
....	11	1011	13
....	12	1100	14
....	13	1101	15
....	14	1110	16
....	15	1111	17
....	16	10000	20
....	17	10001	21
....	18	10010	22
....	19	10011	23
....	20	10100	24

Because the principles for generating the count sequence in each base system are the same, introducing additional base systems should not be difficult once a child appreciates how to count in base two. For example, in base eight, there are eight building blocks: 0, 1, 2, 3, 4, 5, 6, 7. Other numbers can be represented by combining the basic building blocks in the next simplest way (see column 3 of Table 12.1).

Mark, a junior high school student, was referred to me because he was experiencing difficulty in math. Mark seemed bewildered about the whole process of converting from base ten to other base systems. He quickly learned to count in base two and then in other base systems such as base eight. As the session was coming to an end, he asked, "In each base system, is the base number [always represented by] 10 [one-zero]?" Mark noticed that in base two, two is represented by 10 (one-zero); in base eight, eight is represented by 10 (one-zero). To explore the extent of this regularity, we generated the count sequence in additional base systems and confirmed his discovery. Given an informal approach to a mathematical topic, this "math-disabled" boy was beginning to think about and explore the topic.

SUMMARY

To read and write multidigit numerals, children must master an increasingly complex set of rules for decoding/encoding relationships. These rules presuppose knowledge of place-value designation names and knowledge of the number sequence. Deficient place-value knowledge can contribute to incomplete or incorrect decoding/encoding rules. This helps to explain the level-by-level learning of these skills and the difficulties many young children have with three-digit numerals. More specifically, many difficulties with basic numeration skills can be traced to not understanding that position or place defines the value of digits in multidigit numerals. It helps to explain, for instance, why children read numerals as they appear (e.g., reading 402 as "forty-two") or write numerals with extra zeros (e.g., writing "one hundred two" as 1002). Mastery of the powerful but abstract place-value concept is only gradually achieved. Children can rather quickly and mechanically learn the names and value designations of the places. Deeper understanding entails grasping that zero can act as a placeholder; 10 ones, tens, and hundreds are equivalent to 1 ten, hundred, and thousand, respectively; there is a repeating pattern to the system; and multidigit numbers can be partitioned in different ways. Though place-value representation is fundamentally different from children's counting-based view of numbers, it is helpful to teach this new and difficult concept by building on children's informal mathematical knowledge. Specifically, counting activities can be exploited to help children recognize base-ten equivalents or view multidigit numerals as composites of ones, tens, hundreds, and so forth. Counting activities can provide comprehensible models of place-value ideas such as zero as a placeholder.

CHAPTER 13

Written and Mental Multidigit Calculation

What kinds of calculating difficulties do children experience and why? What causes these computing difficulties? Why is encouraging mental arithmetic as important an objective today as ever? What are the psychological prerequisites of mental calculation and estimation? How can teachers build on informal knowledge to minimize or remedy calculating difficulties? How can instruction encourage mental arithmetic skills?

ADAM'S CALCULATING DIFFICULTIES

At 11 years of age, Adam still had difficulty with renaming (carrying and borrowing) procedures. For example, though he had learned a borrowing procedure, he was unsure when to borrow and sometimes borrowed when it was inappropriate. Occasionally he borrowed when given an addition problem. In the problem depicted in Frame A of Figure 13.1, he proceeded to borrow despite the fact that the minuend in the ones place was already larger than the subtrahend! He then laboriously calculated the difference of $19 - 5$ by using marks and a separating-from procedure. He recorded the difference by writing down a 4 below the ones digits and carrying a 1 (not shown). He then subtracted $1 - 1$ for an answer of 0 in the tens place. With larger problems, especially, his work was riddled with errors. For the problem pictured in Frame B of Figure 13.1, he abandoned any effort to borrow and resorted to two incorrect procedures: a small-from-large bug for the ones place and a $0 - N = 0$ (zero-minus-a-number-equals-zero) bug for the tens place.

Adam's mental multidigit calculating ability was also quite limited. When given a problem, such as $32 + 19$, he would try to mentally align the digits in columns so that he could compute a sum for the ones place, tens place, and so forth by counting on. Typically no effort was made to carry (e.g., for $32 + 19$, he answered, "Forty-one"). Indeed, except for the known

FIGURE 13.1: Adam's Written Subtraction Work

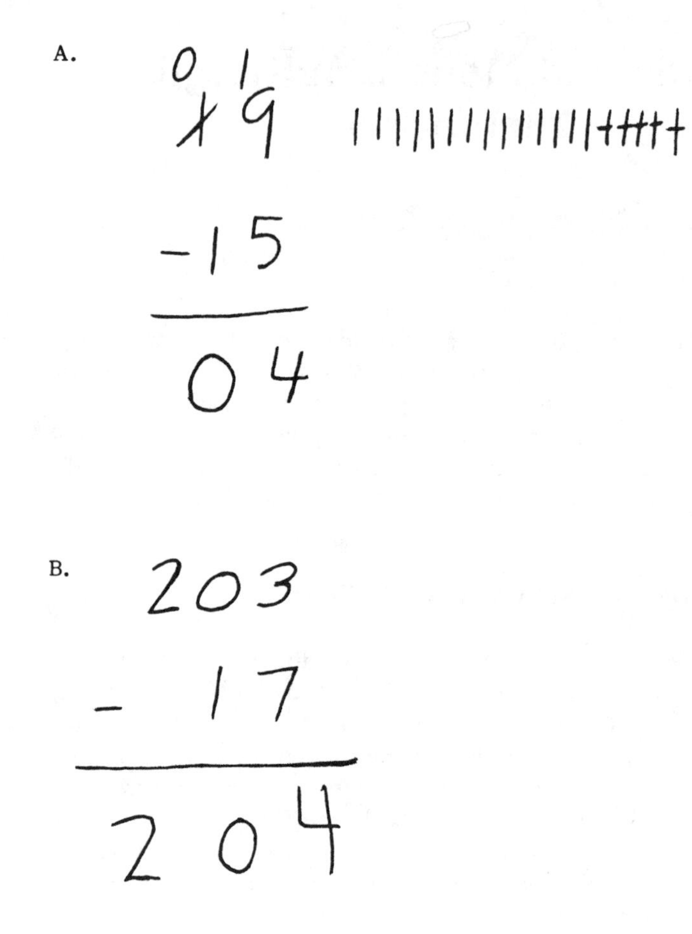

combination 10 + 10 = 20, even the addition of 10 to two-digit numbers (e.g., "32 + 10"), including other decades (e.g., "30 + 10"), was computed by counting on. Subtraction of 10 was laboriously computed by counting down.

Moreover, Adam had great difficulty making estimates. To help overcome his reluctance to make estimates, Adam was introduced to the microcomputer. Adam quickly learned how to enter addition problems into the computer. Before he was permitted to hit the "return" key to see the answer displayed on a monitor, he was asked to estimate the answer. Adam made up the first problem, 356 + 896.

INTERVIEWER: What do you think it will be?

ADAM: It will have to be 12 and some other numbers. [The "12" referred to the sum of 6 + 6 in the ones column.]

INTERVIEWER: How many?

ADAM: 4 [Adam apparently realized that the sum of the tens and hundreds place would each produce two-digit sums.]

Adam made up another problem, 56 + 7093, but again could not estimate the sum. To employ the standard addition algorithm, he attempted to mentally align the numerals but failed. With 57 + 25, he made no attempt to estimate the answer but simply proceeded to calculate the sum mentally. However, by the time he finished the mental calculation, Adam had forgotten the noncarried term in the ones place and so announced, "80 and some number." Even with encouragement to make estimates, Adam much preferred to give exact answers rather than engage in the inexact process of estimating.

In general, Adam had very little sense of arithmetic with multidigit numbers. What accounts for Adam's tremendous inconsistencies in using a borrowing procedure and his incorrect procedures? Why was he unable to mentally add even 30 + 10 or 32 + 10? What accounts for his total inability to make estimates and his tremendous reluctance to even try estimating sums?

WRITTEN CALCULATION

Procedural Rules

Accurate written calculation with multidigit numerals depends on following precisely a sequence of steps (procedural rules). Vertical problems require an alignment rule: Align terms on the right so that ones-place digits form a single column, tens-place digits are lined vertically, and so forth.

For both vertical and horizontal problems, the standard procedure is to start with the ones place and perform the operation indicated on the ones digit. If the problem does not entail renaming, the operation is simply repeated with the digits in the next place to the left (the tens-place digits). This "repeat rule" applies (to the hundreds place, thousands place, etc.) until there are no more digits left in the problem. When terms do not have the same number of digits, a child has to realize that a number plus (minus) nothing is the same as a number plus (minus) zero. For example, in the problem below, the repeat rule does apply to the hundreds and thousands place and, in the hundreds place, 3 + nothing is the same as 3 + 0 or 3:

$$\begin{array}{r} 1342 \\ +\quad 56 \\ \hline 1398 \end{array}$$

As long as the problems do not involve renaming, these few rules can be used to add or subtract terms of even mammoth size, such as:

$$\begin{array}{r} 5237814596 \\ +\quad 2153402 \\ \hline \end{array}$$

The additional rules governing renaming—especially borrowing that involves a zero—may not be plain to many children but, once mastered, tremendously extend their computational reach. For example, the rules for carrying include which digit to carry, where to put it, and what to do with it next. By mastering the rules governing carrying with two-digit numerals, a child, in principle, is capable of adding *any* multidigit numerals. The rules governing borrowing are more intricate than those governing carrying. Embedded zeros present special cases that require additional steps. For example, to borrow from the tens place in the second problem below, a child has to proceed to and borrow from the hundreds place. After making the proper notation, the child can then borrow from the tens place.

$$\begin{array}{rccc} & & 3 & 13 \\ & 7 & \cancel{4} & \cancel{3} \\ - & & 2 & 8 \\ \hline & 7 & 1 & 5 \end{array} \qquad \begin{array}{rccc} & & 9 & \\ & 6 & \cancel{10} & 13 \\ & \cancel{7} & \cancel{0} & \cancel{3} \\ - & & 2 & 8 \\ \hline & 6 & 7 & 5 \end{array}$$

Once children have mastered the rules governing borrowing with two- and three-digit numerals, they only have to learn the procedure for handling back-to-back zeros in the minuend (e.g., 2,003 – 36) to cope with larger problems.

Difficulties Caused by Gaps

Children can and often do learn the procedural rules for multidigit computation and renaming without understanding the underlying base-ten/place-value rationale for such rules. This gap between procedures and understanding often leads to the following types of difficulties:

1. *Alignment difficulties consist of incorrect or inconsistent positioning of digits.* If children do not understand that vertical problems must be aligned from the right, they will probably experience difficulty in situations where they must set up the problem (e.g., with copying problems from a textbook, verbally presented problems, or word problems).

2. *Systematic errors result from incorrect, partially correct, or invented procedures* (as noted in Chapter 4). Typically, bugs are not the result of thinking or memory difficulties but stem from not understanding the underlying rationale for an algorithm (Allardice & Ginsburg, 1983).

3. *Inconsistencies include using a procedure correctly on one occasion but not others.* When procedures are not meaningful to children, they may be unsure when to use a procedure.

MECHANICAL USE OF ROTELY LEARNED PROCEDURES. Procedural rules that are poorly understood are sometimes overextended, too narrowly applied, or haphazardly used. Children sometimes misapply or overextend rules that they learn but do not fully understand. Overextending a rule leads to systematic errors. For example, students are sometimes told that, when doing $N - 0 = \_$ problems, such as $7 - 0 = \_$, "Zero is nothing, so the number (amount) you started with remains the same." Later, some children misapply this rule to the subtraction of $0 - N$ (Kulm, 1985), as in the problem below:

$$\begin{array}{r} 305 \\ -\ 74 \\ \hline 371 \end{array}$$

Children generally are very good at learning and following procedural rules. When it is done blindly, it can lead to inappropriate applications and bugs.

On the other hand, children often use rotely memorized procedures in an overly restricted manner. That is, rotely learned rules may not transfer. Children with learning difficulties, especially, may use procedures correctly only when problems are in a familiar form. If the form of the problem is changed—even slightly—some children fail to see any connection with a known procedure. This may explain why a child is successful on practiced problems (e.g., worksheet assignments of column addition) but not on novel assignments (e.g., word problems or horizontal addition, such as $85 - 39$).

In particular, it is not uncommon for children to accurately calculate with familiar numbers but not apply their procedural rules to larger, unfamiliar numbers. The case of Jenny, a third grader, illustrates this inconsistency. As the problems below show, Jenny had no difficulty borrowing with two-digit terms but failed to apply her procedural rules successfully to problems with three digits. For the first three-digit problem, she borrowed from the tens place, as she had for the first two-digit problem. But then she mechanically and incorrectly proceeded to reduce the hundreds-place term. For the second three-digit problem, she implemented only the first part of the renaming algorithm.

$$
\begin{array}{rr} 3 & 15 \\ \not{4} & \not{5} \\ -1 & 7 \\ \hline 2 & 8 \end{array}
\qquad
\begin{array}{rr} 5 & 10 \\ \not{6} & \not{0} \\ -2 & 4 \\ \hline 3 & 6 \end{array}
\qquad
\begin{array}{rrr} 1 & 5 & 17 \\ \not{2} & \not{6} & \not{7} \\ -1 & 0 & 9 \\ \hline & 5 & 8 \end{array}
\qquad
\begin{array}{rrr} & 10 & 16 \\ 4 & \not{0} & \not{6} \\ - & 7 & 9 \\ \hline 4 & 3 & 7 \end{array}
$$

If children do not understand the underlying rationale for a procedure, they may have difficulty recognizing when to use the procedure even when problems are in a familiar form. As a result, rotely learned procedures may not be used consistently or appropriately. For example, Adam was given the problem $\begin{smallmatrix} & 2\,7 \\ + & 1\,4 \end{smallmatrix}$. Unsure how to proceed, he asked: "Is this the one where I cross out [borrow]?" Because he did not really understand the carrying and borrowing procedures, he was unsure when to use them.

FAILURE TO LEARN PROCEDURES THAT LACK MEANING. Children may not bother to learn or remember information that is meaningless to them. Because they do know the school-taught procedure, they may resort to various strategies for completing assignments. Children who cannot remember (or feel unsure about) the school-taught procedure often fall back on a familiar but inefficient procedure. For example, some children will continue to rely on their informal procedures for problems with a two-digit term. Given a problem such as 23 + 15, the child may count on: "23, 24, 25, . . . , 38."

Quite often, children mechanically and inappropriately apply previously learned procedures to new problems. This is especially likely when the new task is not meaningful to the child. Without a conceptual basis for understanding and learning the new and appropriate procedure, the child may fall back on a well-learned but inappropriate procedure to manufacture some answer. This might help to account for some systematic errors, such as using an incorrect operation (e.g., using familiar addition procedures instead of subtracting) or the small-from-large bug (e.g., 53 – 27 = 34).

Children who cannot remember the correct procedure may resort to inventing their own procedure. For example, like Adam with 203 – 17, Debbie, a second grader, used an incorrect $0 - N = 0$ rule to solve the following problem:

$$\begin{array}{r} 40 \\ -\ 12 \\ \hline 30 \end{array}$$

The use of invented but incorrect rules is common.

INCOMPLETE OR INCORRECT MEMORIZATION. Procedural rules that are not understood are sometimes only partially remembered or remembered incorrectly, and this fact accounts for many difficulties.

When they do not understand the underlying place-value rationale of the procedure, children may be inconsistent in how they align the digits of terms or systematically align digits incorrectly. A child may remember that the digits must be lined up but, without the guidance of place-value knowledge, may not remember whether the digits should be aligned on the left or right. The resulting confusion may cause a child to align terms inconsistently. Because reading requires left-to-right processing, some children may remember the alignment rule incorrectly and, as a result, systematically align the digits on the left.

A child who does not understand the underlying rationale of a procedure may forget steps in the procedure. A partially correct procedure can produce systematic errors. For instance, Kevin, a second grader, could not remember all the steps of the borrowing algorithm. He would borrow but not follow through and reduce the tens-place digit. This incomplete procedure resulted in errors like the following:

$$\begin{array}{r} 40 \\ -\ 12 \\ \hline 38 \end{array} \qquad \begin{array}{r} 44 \\ -\ 36 \\ \hline 18 \end{array} \qquad 42 - 17 = 35$$

Sometimes inadequately understood procedures are learned incorrectly. An incorrectly learned procedure can lead to systematic errors. Boris, a 9-year-old, incorrectly learned the rule for carrying. He consistently placed the carried digit at the top of the left-hand-most column instead of the next column to the left. As a result, he correctly calculated with two- but not three-digit terms:

$$\begin{array}{r} 1 \\ 35 \\ +\ 28 \\ \hline 63 \end{array} \qquad \begin{array}{r} 1 \\ 57 \\ +\ 46 \\ \hline 103 \end{array} \qquad \begin{array}{r} 1 \\ 108 \\ +\ 364 \\ \hline 562 \end{array} \qquad \begin{array}{r} 2\not{1} \\ 168 \\ +\ 156 \\ \hline 414 \end{array}$$

Quite frequently, children's partially correct or incorrect procedures place them in a position where they must invent a new step (Van Lehn, 1983). Consider Boris's incorrect procedure of placing carried terms to the far-left-most column. In the case of the fourth problem, he was confronted with two carried ones. Boris dealt with this perhaps unexpected turn of events by inventing a novel procedure: Add the second carried 1 to the first to make an entry of 2. Incorrectly learned procedures may force a child to invent additional incorrect procedures to "repair" problems caused by the initial bug (Van Lehn, 1983).

Another case in point is Roger, a mildly mentally handicapped adolescent. As Figure 13.2 shows, Roger knew that he had to borrow to subtract the digits in the ones place. Because he could not remember the rules governing borrowing from zero, he skipped over the zero in the tens place and borrowed directly from the hundreds place. Afterward, Roger proceeded as well as he knew how. He treated $0 - N$ as a special case and employed his own $0 - N = N$ rule. To cope with situations for which they have only incomplete procedural knowledge, children many times combine their own invented procedures with partially correct procedures.

Bug Stability

How consistent are children's systematic errors? If they lack conceptual knowledge, children may deal with a new task in terms of their existing or available procedural knowledge for some time. Thus children may use inappropriate, partially correct, or invented procedures on a regular basis for an extended period of time. In such cases, the resulting systematic errors remain consistent or stable over a relatively long period of time. Sometimes children use inappropriate, partially correct, or invented procedures as a temporary expedient for a task they do not understand. When they encounter a similar situation (a few problems or days later), children may not use the same stopgap measure but invent a new makeshift device. In these cases, errors are "systematic" but not stable. Such bug instability may be especially prevalent when children are first introduced to a new procedure and have not yet fixed on one way of coping with the novel task.

Sometimes bug instability takes another form: A child may use the correct

FIGURE 13.2: Roger's Combination of Bugs

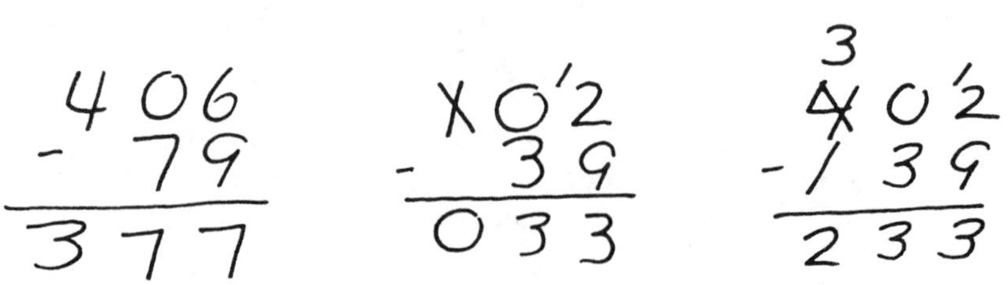

procedure and a systematic bug intermittently. This inconsistency has several causes. First, if a child does not understand the rationale for a procedure, the child may be uncertain about which procedure to use. As a result, the child may mechanically use correct and incorrect procedures interchangeably. Second, a child can know the correct procedure but, because of affective factors, may slip into an incorrect but systematic solution method. That is, under some circumstances, a child might be highly motivated to use the correct procedure. Under conditions of low motivation, a child might resort to buggy procedures to save effort. For example, with a problem such as 206 – 77 = __, the correct renaming algorithm would require more effort than employing the small-from-large buggy procedure to answer 271.

In conclusion, stable bugs reflect not only a problem with procedural knowledge but also a deficiency in conceptual knowledge—a gap between instruction and understanding. Unstable bugs may also reflect deficiencies in procedural and conceptual knowledge. However, slipping back and fourth between correct and systematically incorrect procedures may sometimes be due to a lack of interest. Whether stable or unstable, systematic errors violate the rationale of the procedure and indicate a need to adjust a child's instruction.

MENTAL ARITHMETIC

Key Skills

Mental arithmetic can involve determining exact answers (mental computation) or making educated guesses (estimation). Mental computation and estimation are indispensable skills (e.g., Bell, 1974). Mental arithmetic can serve to foster quantitative thinking, check written calculations, and solve numerous everyday problems.

FOSTERING QUANTITATIVE THINKING. Mental arithmetic can lead to the discovery of the patterns, properties, and structure of our number system (Reys, 1984). For example, like Aaron, described in Chapter 7 ("1,000 plus 1,000 is . . . 2,000"), many children informally learn that the relationship "one plus one is two" applies to larger combinations, such as *one* hundred plus *one* hundred is *two* hundred, and even *one* billion plus *one* billion is *two* billion. This connection may serve as the basis for discovering that the number system is full of recurrent arithmetic patterns (e.g., 4 + 3 = 7, 40 + 30 = 70, 400 + 300 = 700, 4,000 + 3,000 = 7,000, and so forth). Such informally discovered relationships can later help inform written calculational efforts with multidigit numerals.

Mental arithmetic also promotes more flexible quantitative thinking. Mental calculation and estimates encourage children to think in chunks of tens and multiples of ten. For instance, it is easier to mentally compute 542 + 135 in terms of hundreds, tens, and ones than it is to use the standard algorithm. The standard right-to-left approach (2 + 5 = 7, 4 + 3 = 7, 5 + 1 = 6: 677) requires storing and then transposing (reversing the order of) the partial sums to state the answer. By using the nonstandard procedure, a child can add from left to right and avoid transposing digits (e.g., "[500 + 100 is] six hundred, [40 + 30 is] seventy, [2 + 5 is] seven"). Indeed, my observations of bright third graders indicate that these children naturally adopt and prefer their own left-to-right approach, even when carrying is involved. Research (e.g., Ginsburg, Posner, & Russell, 1981) suggests that a regrouping strategy (e.g., 57 + 26 = ([50 + 20] + [7 + 6]) is widely adopted: from older schoolchildren and adults in the United States to unschooled West African merchants.

Because it promotes understanding and flexibility, the encouragement of mental arithmetic may have some very important side effects. Children may gain a sense of command over large numbers (Trafton, 1978). Moreover, some research (e.g., Driscoll, 1981) indicates that mental-computation training can improve children's general mathematical problem-solving performance (Reys, 1984).

CHECKING. Mental calculations are often used by adults to check their own written calculations, the calculations of others, and machine-printed calculations. Estimation is particularly useful in quickly checking the reasonableness of calculated answers. With the popular use of computers and hand-held calculators, estimation has become even more important as a checking skill (e.g., Hope, 1986; Trafton, 1978). For instance, in tabulating the costs of four items ($1,027.99, $396.00, $2,333.67, and $27.95) on a calculator, a buyer finds that the total comes to $6,552.66. Is this reasonable? The buyer quickly makes an estimate using the most expensive items: $1,000 + $2,000 is $3,000 + a few hundred. Even with the remaining items, the cost should be no more than $4,000. (The cost was calculated incorrectly because $27.95 was punched in as $2,795!)

APPLIED PROBLEM SOLVING. Mental computation and estimation are also an integral part of much everyday problem solving. For example, to figure out a 15% tip for a $32 meal, a polite diner does not take out a pen and paper or a calculator. Instead, the diner mentally calculates the tip. A favorite procedure involves taking 10% of the bill and adding another half (5%): $3.20 + $1.60 = $4.80.

Estimation is important in everyday situations where exact answers are not necessary, difficult to obtain, or even impossible to calculate (Trafton, 1978). Indeed, most of the mathematics used by adults involves estimation (Reys, 1984). For example, to stay within budget, a grocery shopper rounds off the cost of each item picked up to keep an estimate of the total cost: \$2.00 (for a gallon of milk at \$2.10 per gallon); \$2.00 + \$1.50 (for a box of cereal at \$1.42); \$3.50 + \$1.50 (for a 2-quart bottle of juice at \$1.39); \$5.00 + \$2.00 (for a 3-pound jar of applesauce at \$1.79), \$7.00 + \$1.00 (for plastic wrap at \$1.19 per roll); and so forth. At the checkout counter, the grocery shopper gives the cashier five \$20 bills for an \$87.66 total and quickly estimates the change that should be returned: "\$87.66, okay, \$88, \$2 to get \$90, another \$10 to get \$100 – I should get a ten, two ones, and a small amount of change." In brief, mental calculation and estimation are used so widely that Trafton (1978) has proposed that mental-arithmetic skills should be a part of the minimal mathematical competency expected of all citizens.

Multidigit Mental Computation

PREREQUISITE FOR MENTAL CALCULATION. Mental arithmetic requires facility with basic *and* large number combinations – particularly combinations with 10 (e.g., $10 + 7 = 17$, $30 + 10 = 40$, $42 + 10 = 52$, $42 - 10 = 32$). Mastery of basic-number combinations is critical for mental calculation for two reasons. First, automatic production of large-number combinations (e.g., $400 + 300 = 700$ or $400 - 300 = 100$) depends on mastery of basic combinations (e.g., $4 + 3 = 7$ and $4 - 3 = 1$).

Second, production of the basic combinations is an integral component part of mental calculation. For example, to use the standard algorithm to mentally sum $432 + 165$, a child must determine the sum of the ones-place term ($2 + 5 = 7$) and keep this partial result in working memory as the partial result of the tens place ($3 + 6 = 9$) is obtained. Then both of these partial sums must be kept in mind as the sum for the hundreds place ($4 + 1 = 5$) is obtained. With addition problems that involve renaming, there is the added difficulty of keeping track of carried values. Mentally using the standard borrowing algorithm is even more complicated because a child must keep track of reduced terms and rename terms. To minimize the load on working memory and the chance of error and confusion, it is essential to determine partial sums (differences) efficiently. Indeed, because each partial result is not recorded, automatic production of the basic number combinations is even more important for mental calculation than it is for written calculation.

Facility with large combinations, especially those involving 10 and later 100 and 1,000, is pivotal for efficient multidigit mental arithmetic. For example, automatically adding 10 can reduce dependence on the laborious counting-on procedure or provide a shortcut for the standard algorithm. When adding 10 and a single-digit number ($10 + N$ or $N + 10$), such as $10 + 7$, automatic recognition that the sum is N + teen (seventeen) can eliminate the need to count on ("10; 11, 12, 13, 14, 15, 16, *17*—17") or proceed through the standard algorithm ("0 and 7 are 7; 1 and nothing are 1: 17"). Children master combinations involving 10 and 100 fairly quickly. Only a modest amount of informal calculation may be needed to discover the relatively straightforward $10 + N$ or $N + 10 = N$ + teen relationship (e.g., 10 + 7 = seven + teen). Children may also quickly see the relationship that adding any (one- or two-digit) number to 100 yields one hundred- + the number (e.g., 100 + 7 = 107, 100 + 30 = 130).

A psychological basis for automatically adding (or subtracting) 10 with decades (e.g., 30 + 10 = 40) and other two-digit numbers (e.g., 32 + 10 = 42) is an ability to count by tens to 100 ("10, 20, 30, . . . , 100"). Once children know the order of the decades, they can automatically cite decade-after (decade-before) relationships (e.g., after 30 comes 40 when we count by tens). Once children know decade-after (decade-before) relationships, they simply have to see the connection between the decade structure (counting by tens) and arithmetic to automatically add (subtract) by 10.

Likewise, counting by tens over 100 provides the basis for adding and subtracting 10 with three-digit numbers. To count by tens over 100, children have to recognize the repetitious nature of the decade sequence. That is, a child has to realize that the same decade sequence from 10 to 90 applies but with the addition of a hundreds marker: one hundred + ten, one hundred + twenty, one hundred + thirty, . . . , one hundred + ninety. Once children recognize this repetitive pattern and master the highly regular hundreds sequence (one hundred, two hundred, three hundred, . . . , nine hundred), they can count by tens all the way to 990.

Estimation

PREREQUISITES FOR ESTIMATION. Children need good mental computational skills in order to develop good estimation skills (Reys, 1984). For example, the ability to realize quickly the effects of adding and subtracting ten and multiples of ten is basic to estimating skills (Payne & Rathmell, 1975). An understanding of place value is also critical for estimation skills (Trafton, 1978).

ESTIMATION STRATEGIES. The most familiar estimation strategy is *rounding* (Trafton, 1978). Estimation by rounding is an example of practical problem solving. The nature of the problem to be solved determines what unit to round to (Reys, 1984). For example, to make the task of mental computing more manageable, the grocery shopper in the example above decided to round to the nearest half dollar. Someone interested in a less precise estimate might have rounded off to the nearest dollar. A shopper on a tighter budget who needed a more precise estimate could round off to the nearest tenth of a dollar (ten cents). Estimation by rounding also entails deciding how to round. A standard procedure is to use the halfway point. In rounding off to dollars, $1.51 to $1.99 is more than halfway and is therefore rounded up to $2; $1.01 to $1.49 is less than halfway and so is rounded down to $1. An amount that is exactly halfway, such as $1.50, is rounded down, in this case to $1. (There are different conventions; the IRS now has taxpayers round up.) In making estimates, however, the standard rounding rules may not be applicable. For example, the grocery shopper who wants to make very sure that the available cash is not overspent may round up. Estimation by rounding, then, requires thoughtful analysis of the problem and flexibility in choosing a solution procedure—hallmarks of problem solving.

There are a number of other useful estimation strategies. Exploiting *compatibles* is similar to but more sophisticated than rounding (Reys, 1984). This strategy involves changing otherwise difficult problems into a form that is compatible with well-known knowledge. Consider the following bill of sale:

$$\begin{array}{r} \$\ 6.85 \\ \$\ 4.99 \\ \$14.75 \\ \$\ 2.92 \end{array}$$

Rounding would produce the following:

$$\begin{array}{r} \$\ 7 \\ \$\ 5 \\ \$15 \\ +\$\ 3 \\ \hline \$30 \end{array}$$

Mental computation could be made even easier by recasting the problem into familiar combinations: ($7 + $3) + ($5 + $15) = $10 + $20 = $30. By converting an unfamiliar into a familiar problem, a child is actively engaged in problem solving.

Averaging is also similar but more sophisticated than rounding. Averaging is useful when numbers cluster around a common value (Reys, 1984), as in the following bill of sale.

$5.42
$4.81
$5.95
$4.02
$5.10

An estimator must decide whether the values do tend to cluster around a particular value and what the value is. In the case above, the values do cluster rather evenly around $5. Thus an estimate can be obtained by counting five, five times (5, 10, 15, 20, 25), or multiplying five times five. Again, this technique involves a thoughtful analysis of the problem.

A *front-end* strategy can be useful when problems include numbers of three or more digits (Trafton, 1978). This approach is illustrated by the buyer in the calculator example discussed earlier:

$1,027.99
396.00
2,333.67
27.95
____

The buyer identified the digits that would have the most significant impact on the outcome (Reys, 1984). Clearly the digits (1 and 2) in the thousands place were the most important in determining the magnitude of the sum. The total of these front-end digits (3) provided a ballpark estimate of the sum ($3,000 +). The advantage of this approach is that it does not involve rounding, nor does it require a deep understanding of base-ten/place-value representation. Because the numbers are visible, even young children can arrive at decent estimates quickly and experience success (Reys, 1984).

Adjusting is a relatively sophisticated estimation process. For example, the front-end approach used by the buyer to check the calculated sum produced a gross estimate of $3,000 + . By checking the next most significant digit (the hundreds digit), the buyer could determine where the cost was between $3,000 and $4,000. That is, the $1,027.99 and $27.95 do not change the original estimate, but $2,333.67 and $396.00 pushed the estimate to $3,600 – still well below the $4,000 mark. Considerable familiarity with and understanding of base-ten/place-value and computation is required to make quick and accurate adjustments – particularly fine adjustments.

Difficulties

Instruction based on absorption theory underplays the importance of mental arithmetic. Textbook series often treat mental computation and estimation as supplementary or secondary topics. Furthermore, standardized tests, which greatly influence what topics are emphasized, usually fail to measure mental arithmetic (Reys, 1984). The result is that mental arithmetic often receives insufficient attention.

MENTAL COMPUTATION. Mental computation difficulties arise when children have not mastered the basic or large-number combinations. If children do not know basic combinations (e.g., $2 + 9 = 11$) but have to figure out partial sums, working memory quickly becomes overloaded. This can result in slips or confusion. For example, when given $32 + 19$ to calculate mentally, Adam had to count on to determine the sum of $2 + 9$ ("9; 10, 11"). The added burden of keeping track of this mental computation may have contributed to his losing track of the carried term and obtaining the wrong answer ($32 + 19 = 41$).

Many children spontaneously discover that basic number combinations, such as $4 + 3 = 7$, reoccur as larger combinations. However some children may need to be guided to this discovery. For example, though he knew $4 + 1 = 5$ and $2 + 2 = 4$ at the time, Adam did not see the connection with larger problems, such as $40 + 10 =$ __, $400 + 100 =$ __, $20 + 20 =$ __, or $200 + 200 =$ __. Furthermore, children like Adam, who have not mastered the basic number combinations, such as $4 + 3 = 7$, are not likely to learn larger combinations such as $40 + 30 = 70$.

Mental computation can also be hindered when children fail to discover straightforward arithmetic relationships involving the addition of 10 and single-digit numbers or 100 and one- or two-digit numbers. For example, Adam dealt with problems such as $10 + 7$ or $100 + 30$ by counting on: "10; 11, 12, 13, 14, 15, 16, 17" and "100; 101, 102, 103, . . . ,128, 129, 130," respectively. This approach is tedious and prone to error. In cases such as Adam, children need help discovering adding-with-ten and adding-with-one-hundred rules.

Difficulties with mentally computing with 10 can occur because children have not developed the prerequisite (count-by-tens and decade-after/-before) skills, or because they fail to see the connection between existing decade knowledge and adding or subtracting with 10. For instance, though Adam could count by tens to 100 with facility, he did not see the connection between this count knowledge and adding 10 to decades. As a result, given $30 + 10$, he counted on. Furthermore, he was unable to count by tens over 100. A direct result was that he had great difficulty mentally adding with a three-digit term. Unfamiliar with the decade sequence over 100, he could not solve problems such as 130 + 10 automatically or count on by tens for problems like 150 + 30 ("150; 160, 170, 180").

ESTIMATION. Children can have difficulty making effective estimates because they lack the (a) prerequisite skills and concepts, (b) estimation strategies themselves, or (c) flexibility to engage in an inexact but thoughtful problem-solving process. As the case of Adam illustrates, children who do not have facility with mental computation will have great difficulty making estimates. Adam's inability to efficiently add or subtract 10 and his inability to think in groups of tens, hundreds, and so forth made estimation an impossible task.

Moreover, Adam had not been taught estimation skills—possible approaches to the task of estimation. Estimation is usually overlooked by primary instruction based on absorption theory. Indeed, the National Assessment of Educational Progress (Carpenter, Coburn, Reys, & Wilson, 1976) concluded that estimation is "one of the most neglected skills in the mathematics curriculum." Basal series devote little attention to developing estimation skill during the primary years (Driscoll, 1981). When given at all, estimation instruction is frequently very limited in scope. Training often focuses on only the strategy of rounding (Trafton, 1978). In addition, training often fails to establish clearly the connection between rounding and estimating (Reys, 1984). That is, instruction mechanically teaches children to round without helping them to understand that rounding is *a* means of making the terms manageable for mental computing. The result is that children often do not have the skills to make good estimates. National surveys suggest that U.S. children do poorly on tests of computational estimation (e.g., Carpenter et al., 1976).

Finally, Adam—like so many children—was reluctant to make estimates. Indeed, many children are uncomfortable with estimation (Trafton, 1986) and respond rigidly to estimation tasks. Many children so value getting the "right answer" that they try to compute an exact answer even when asked to make estimates. Teachers frequently admonish children not to respond quickly without thinking (in itself good advice) with, "Don't just guess" (Carter, 1986). Unfortunately, children take these words to mean that giving inexact answers (estimation) is not acceptable. Moreover, children like Adam often have little confidence in their estimation skills. Afraid of ridicule, they pursue computing exact answers.

EDUCATIONAL IMPLICATIONS: PROVIDING MEANINGFUL AND WELL-ROUNDED COMPUTATIONAL TRAINING

When mathematical tasks are imposed upon children without regard for their understanding and interest, children quite naturally tend to deal with the tasks in a mechanical fashion. Below are suggestions for introducing, practicing, and reteaching computational skills in a more meaningful and effective manner.

1. *Introduce renaming procedures informally with concrete models.* The base-ten/place-value rationale for the renaming procedures is a rather abstract and unfamiliar notion to children. And all the procedural rules for renaming are not easily memorized without understanding. Fortunately, like other formal knowledge, the rationale and procedures for renaming can be taught to children by building on their informal knowledge. Learning the procedural rules in a meaningful manner can minimize many of the difficulties described above.

Forward Bowling (Wynroth, 1969/1980), described in Chapter 12 as a means of teaching base-ten/place-value concepts and skills, also serves as a concrete model for the carrying procedure. The rules of the game specify that no more than nine chips can remain on the ones-place peg, tens-place peg, and so forth. If a player adds, say, 5 chips to 7 already on the ones peg, the rule requires the child to reduce the number of chips to 9 or less. Another rule of the game specifies how this is done: 10 chips on a peg can be traded for 1 chip that is put in the next peg. Hence the child would remove 10 of 12 chips on the ones peg and replace them with 1 chip on the tens peg. This concrete procedure corresponds to the written procedure of adding the ones digit 7 + 5 to produce 12, renaming the 12 ones as 1 ten and 2 ones, and entering the 1 ten in its own (the next) column.

Backward Bowling (Wynroth, 1969/1980) serves as a concrete analogy for the borrowing procedure. In backward bowling, players start at a predesignated score (say 199) and remove chips for each pin knocked down (see Figure 13.3). The first player or team of players to remove all their chips is the winner. Note that this game, like forward bowling, provides a concrete model of zero as a placeholder (e.g., see Step D in Figure 13.3).

Addition with hieroglyphics was introduced to Adam—with simple problems (e.g., I plus I, IIIIIII plus II, ∩ plus III) followed by more complex problems (e.g., IIIII plus IIIII). He handled all of these problems quite capably. Note how nicely the Egyptian system meshes with children's informal counting approach to addition. Later he was asked to translate problems such as 53 + 27 into hieroglyphics (∩∩∩∩∩III + ∩∩IIIIIII) and give the

FIGURE 13.3: Scoring Procedure for Backward Bowling*

Situation: Child starts turn with a score of 114 points and then knocks down eight pins.

Step A: Count the number of pins knocked down.

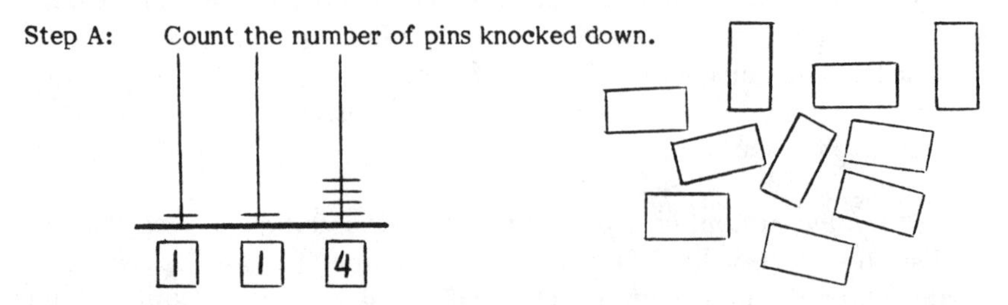

Step B: From the one peg, remove blocks equivalent to the number of pins knocked down. If there is an insufficient number of blocks on the one peg, trade in a "ten block" for ten ones.

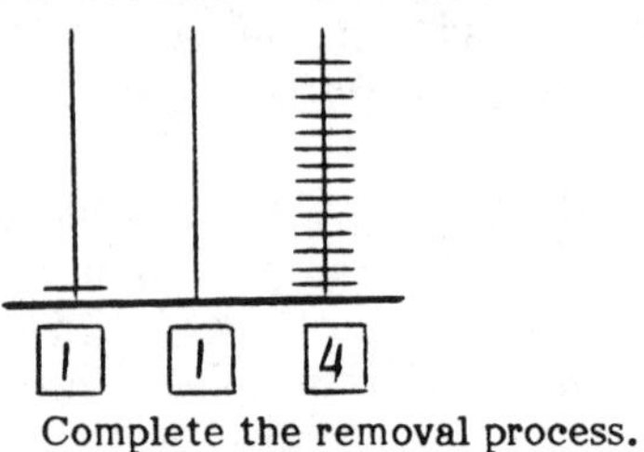

Step C: Complete the removal process.

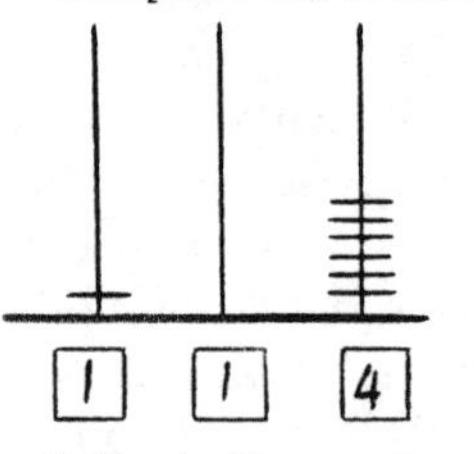

Step D: Indicate the number of blocks on each peg with numerals and read.

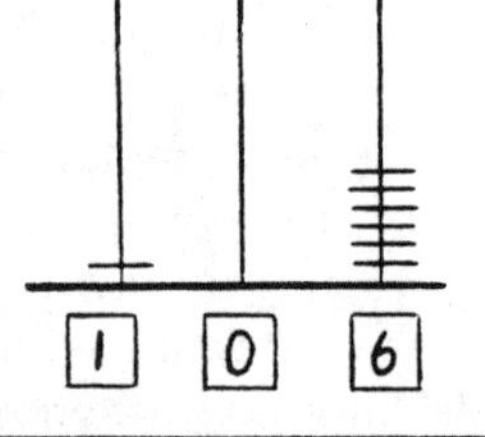

*Based on Wynroth (1969/1980).

answer in both the Egyptian and our number system (∩∩∩∩∩∩∩∩ and 80, respectively). Working out this problem in Egyptian hieroglyphics provides a concrete model for the renaming concept. After noting that there were a total of 10 unit symbols, Adam equated this partial sum with *a* group of ten: He wrote down a ∩ that he then added to the other hieroglyphic symbols for ten.

2. *Explicitly link formal procedures to concrete models.* Children may not spontaneously see a connection between concrete models for base-ten/place-value ideas and written renaming procedures (Resnick & Ford, 1981). Therefore, teachers should introduce renaming algorithms in a manner that makes the link to concrete models clear. For example, it might be helpful to introduce the carrying algorithm by showing how it parallels adding with the more concrete hieroglyphics. In a problem such as $\begin{array}{r} 27 \\ +18 \\ \hline \end{array}$ point out that trading 10 ones (I) for a ten symbol (∩) in hieroglyphics is represented in our system by breaking up the sum of the ones place (15) into a ten (the one that is carried to the tens place) and five ones (which are recorded in the ones column).

The Wynroth (1969/1980) curriculum provides an excellent illustration of how to make the links among concrete models, concepts, and procedures explicit. Children are introduced to a renaming notion by means of the forward- and backward-bowling games described above. These concrete renaming procedures are introduced in the first grade—long before they are introduced to the written arithmetic with renaming. After the concrete scoring procedures are mastered, a teacher points out that written arithmetic is the same thing the children did to keep track of their bowling scores, except that the written work uses numbers. Indeed, the written procedure can be introduced in terms that parallel the scoring procedures the children learned playing the bowling games (e.g., see Figure 13.4). Moreover, at least initially, students are encouraged to represent and compute the answers to symbolic problems using the blocks and the pegboard used for the bowling games. In this way, the previous, counting-based instruction is explicitly linked to the formal arithmetic it is intended to inform.

3. *Practice algorithms in a meaningful and interesting manner.* Children sometimes slip into buggy procedures or fail to complete assignments because they grow tired of dull and useless tasks. Moreover, when a task seems pointless to children, sometimes no amount of practice will promote accurate retention. Children, like adults, are more prone to practice and master a skill if the skill is personally important to them. Keeping score for games such as those described above can be personally important to children. In such a context, learning to compute and computing are viewed as a necessary aspect of an engrossing activity. Thus scoring games can pro-

FIGURE 13.4: Explicitly Linking the Carrying Procedure to a Concrete Scoring Procedure

Step 1: Add the ones-place terms.

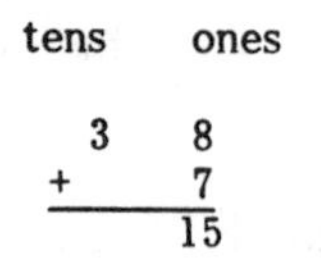

Step 2: Make sure the entry in the ones place is legal: As in the scoring procedure for forward bowling, a number no larger than 9 (only a one-digit term 0 to 9) is allowed.

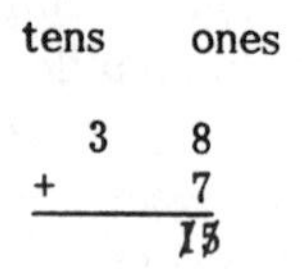

Step 3: Because there are more than 9 in the ones place, 10 ones must be traded in for 1 ten.

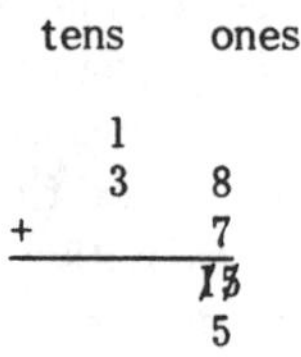

Step 4: Use the same procedure with the tens place.

tens ones

1
3 8
+ 7
15
4 5

vide considerable practice that does not seem like pointless drill to children. Practice can also be made interesting for many children by using microcomputers.

4. *Encourage checking written calculations against answers derived by informal procedures.* Because children are familiar and comfortable with informal arithmetic procedures, they trust such procedures. When they are first mastering a formal algorithm, it may be helpful to encourage them to compare the results of their school-taught procedure against their infor-

mally derived solutions. This gives children their own means of determining whether the solution is sensible or not.

Indeed, by contrasting the formally and informally derived solutions, it may help the child to "troubleshoot" the procedural errors in his school-taught algorithm. Adam, like many children with math learning problems, had great difficulty with written addition involving carrying. Consider the problem below:

$$\begin{array}{r} 37 \\ +8 \\ \hline \end{array}$$

At various times he responded by writing 315, 15, or nothing at all. Asked to solve such problems by counting, Adam had no difficulty in employing his counting-on procedure (37, 38, 39, 40, 41, 42, 43, 44, 45) to produce the correct answer. To help Adam connect the unfamiliar written procedure (e.g., 37 + 8: 7 + 8 is 15, carry the 1, 1 + 3 is 4; 45) with his familiar informal strategy, I had him first compute the answer by counting on, record the answer, and then use the written algorithm. In this way, Adam had a sensible basis for checking and correcting his written procedure. When a written answer did not match his counted answer, Adam would try to modify the written procedure to produce the right answer. Juxtaposing a familiar and reliable informal strategy with a formal procedure may make learning the latter less threatening and more meaningful.

5. *Remedial efforts should focus on encouraging understanding as well as learning the correct procedure.* Sometimes remedial efforts that focus only on reteaching the faulty step(s) in an algorithm will be successful. Sometimes, however, a teacher may help a child to relearn a particular step only to find that later the child exhibits another difficulty. For example, a new bug appears. This often happens when the child does not really understand the underlying rationale for an algorithm. As a result, the child keeps running into difficulty (e.g., forgetting some aspect of the procedure). Instead of attempting a quick fix of the child's procedural knowledge, remedial efforts should reteach the skill in a meaningful manner as described above. This may take more time and effort but, in the long run, may be less frustrating for teacher and child.

Mental Arithmetic

1. *Mental computation and estimation should be integral aspects of the primary curriculum.* Mental computation and estimation are basic survival skills, and their use helps promote mathematical thought. Therefore, these skills should be encouraged from the beginning of formal school.

Postponing or shortchanging instruction of these skills encourages unwarranted biases that may be very difficult to overcome or remedy later.

2. *Foster an appreciation for mental calculation and estimation.* It is essential that students understand the value of these skills (Trafton, 1986). Practicing and testing these skills is concrete evidence of their importance. Discussing the purposes or uses of these skills may help to underscore their value. For example, point out that the reasonableness of answers generated by a calculator or computer needs to be checked. Give children exercises where they have to "troubleshoot" computing errors: A football team scored 21 points the first period, 7 the second, 0 the third, and 3 the fourth. However, the coach's calculator showed a total of 11 points at the end of the game. What did the coach do wrong?

3. *Ensure mastery of the psychological prerequisites for mental arithmetic.* Efficient mental calculations require facility with basic number combinations and large number combinations, especially those involving 10 and 100. Mastery of the latter combinations depends on seeing recurrent patterns in the number sequence and facility with the decade sequence. Efficient mental calculation, especially, with groups of 10, 100, and so forth, is crucial for estimation. Deficiencies in prerequisite skills must be remedied first. One method for developing mental addition and subtraction of 10 is described in Example 13.1.

4. *Encourage children to look for ways to shortcut mental arithmetic.* Children should be explicitly encouraged to look for ways to make mental calculation and estimation easier. This exploration can lead to the discovery of important relationships. Have a child describe an invented mental procedure to others. This will be rewarding for the inventor and may be helpful to other children.

5. *Estimation instruction should focus on teaching a variety of specific strategies.* Estimation should not focus exclusively on rounding. Children should be shown various other approaches, such as a front-end strategy (see, e.g., Leutzinger, Rathmell, & Urbatsch, 1986; Reys, 1986).

6. *Point out the rationale for mental arithmetic procedures.* It is not enough, for example, to teach estimation strategies. Children may need help understanding the rationale behind the procedures. It may be necessary to help children see that rounding is *a* means of making the numbers more manageable for mental computation.

7. *Children should be encouraged to think of estimation as a problem-solving process.* Estimation is not merely a skill that children can learn by rote (e.g., Trafton, 1986). It is a set of skills that must be applied thoughtfully to accomplish particular objectives. Children may need help deciding when estimation would be useful or which estimation strategy would be most suited for the task. They need experience analyzing problems and

EXAMPLE 13.1

CARD GAME "99"

Objective: Mental addition and subtraction of 10

Materials: Standard deck of cards or deck of cards with printed instructions (see Procedure section)

Procedure: Briefly, the object of play is to avoid discarding a card that puts the discard total over 99. Three cards are dealt to each player. Cards have their face value except:

Aces (which adds 1 or 11 to the discard total);

4 (adds 0, but reverses the direction of play);

9 (automatically makes the discard total 99);

10 (deducts 10 from the discard total);

Jacks (adds nothing);

Kings or queens (adds 10).

Instead of using a standard deck, it may be less confusing to some children to make up a deck with only the relevant instructions: + 1 or + 11, Reverse, 99, -10, + 0, and + 10. The player to the left of the dealer starts by discarding, announcing the new total, and drawing a replacement from the deck. Play continues until someone loses (puts the discard total over 99). The loser gets one strike. The game ends when one player accumulates three strikes. The player or players with the fewest number of strikes win(s).

deciding how much rounding is required to accomplish an intended objective.

8. *Use calculators and computers.* Calculators and computers can be very helpful in developing mental arithmetic skills. Using these tools can provide a strong incentive for developing good estimation skills (Reys, 1984). Play estimation games in which children try to estimate answers and then use calculators to check.

SUMMARY

A gap between computational instruction and understanding frequently surfaces in the form of alignment difficulties, systematic errors, or inconsistencies. These difficulties arise because rotely learned procedures are frequently used without careful thought or reflection. Indeed, a rote-mastery approach often fails to promote accurate learning of procedural rules and sometimes produces no learning at all. Meaningful instruction of written calculation algorithms builds on children's informal and concrete arithmetic. Games provide a natural and motivating means to teach renaming concepts and practice computational skills. Mental arithmetic is important because it encourages children to think and solve problems more flexibly and is a widely used everyday skill. Indeed, in the age of the computer, estimation is crucial as a checking skill. Mental arithmetic depends on a ready facility with basic and multidigit number combinations and place-value concepts. It is essential that formal schooling foster these basic prerequisites or encourage children to develop and practice mental arithmetic skills.

CHAPTER 14

Problem Solving

Before formal instruction, are young children capable of solving arithmetic word problems? How do U.S. children fare on problem-solving tests? What is required for successful problem solving? How do textbook problem-solving exercises compare to everyday or mathematical problem solving? How can problem-solving competence be encouraged?

EARLY PROBLEM-SOLVING ABILITY

Adam's Problem Solving

Adam answered the written problem $\begin{array}{r} 66 \\ +\ \ 4 \\ \hline \end{array}$ with 610. Because the 11-year-old learning-disabled boy had not mastered the renaming algorithm, he did not carry but simply recorded the sums for the ones and tens place below the line. Given a comparable verbal word problem ("If you have 66 stamps in your stamp book [Adam was a stamp collector] and you put in 4 more, how many stamps does the stamp book contain now?"), Adam proceeded to count ("66; 67, 68, 69, 70") and announced the correct answer: "70."

A year later, Adam was asked to determine the difference between two scores: "You got 30 points and I got 8. By how many points did you beat me?" Adam attempted to count up from 8: He wrote 9, 10, 11, . . . , 30. Because he did not keep track of or count how many numerals he wrote down, Adam did not respond correctly. Instead he simply announced the cardinal name of the last numeral recorded: "30." After it was pointed out that 30 was his score and could not be the difference between our scores, Adam contemplated the problem and revised his strategy. This time he verbally counted up from 8 and made marks as he counted. After he arrived at "30," he counted the marks to obtain the difference: "22." Adam was then asked if written subtraction was applicable to the problem.

INTERVIEWER: Would that [writing out and pointing to the written problem depicted below] give the right answer?

$$\begin{array}{r} 30 \\ -\ 8 \\ \hline \end{array}$$

ADAM: No.

INTERVIEWER: Your score is 30 and my score is 8. . . . Will this tell us the difference between our scores?

ADAM: [Apparently confused.] It's eight.

In succeeding trials and sessions, Adam persisted in using counting up to compute differences. Moreoever, he resisted the idea that written subtraction or other informal subtraction procedures, such as separating from or counting down, were applicable to difference problems.

Even though he could not solve the written problem 66 + 4, Adam's informal arithmetic knowledge enabled him to understand and solve simple addition word problems. Is it typical that children can solve verbal word problems even before they learn formal arithmetic? Note that, with a minimum of guidance, Adam was able to construct an informal but accurate solution strategy for difference problems. Do young children typically construct their own counting strategies when introduced to word problems? Adam seemed to interpret difference problems as a unique kind of problem, as distinct from other kinds of subtraction problems such as "take-away." As a result, he saw only one way that such difference problems could be figured out: by counting up. Do children typically reserve a particular kind of strategy for a particular type of word problem? Are young children's choices of solution strategy for a particular type of word problem usually so inflexible?

Competence Before Formal Instruction

It is often assumed that young children are not capable of analyzing and solving arithmetic word problems before formal schooling. Indeed, it is commonly believed that word problems are a relatively difficult task and that their introduction should be delayed until *after* children have mastered basic formal arithmetic skills. Contrary to popular notions, research (e.g., Carpenter, Hiebert, & Moser, 1981; Carpenter & Moser, 1982, 1983, 1984; Court, 1920) indicates that before receiving formal arithmetic instruction, most young children can use their informal arithmetic knowledge to analyze and solve simple addition and subtraction word problems (Carpenter, in press).

In fact, as in the case of Adam, simple arithmetic word problems are frequently more meaningful to young children than formal or symbolic arithmetic. Mark, a kindergartner, was shown the written addition problem 5 + 1 and asked how much five and one were altogether. The child looked perplexed and did not respond. When the symbolic problem was put in terms of a word problem ("How much are five candies and one more candy?"), the child thought a moment and responded, "Oh, six." The child then responded correctly to the remaining five symbolic problems.

It is not uncommon for children to respond blankly at first to symbolic arithmetic problems. Unable to interpret the foreign symbols 5 + 1, Mark had no idea how to respond. However, the word problem concerning candy could be interpreted in terms of his informal understanding of addition as an incrementing process and hence was meaningful. Because the word problem could be assimilated to the child's informal knowledge of arithmetic, he then could devise or recreate a strategy to determine the sum. By connecting the foreign symbolic problem to a meaningful word problem, the formal arithmetic became understandable. Thus Mark was able to solve the remaining symbolic problems without further help.

Informal Solution Methods

When given arithmetic word problems, Adam relied on informal solution procedures. Furthermore, he appeared to interpret different types of problems in terms of different informal arithmetic concepts and thus selectively applied different counting-based procedures. Specifically, he interpreted the stamp problem (66 stamps and 4 more) in terms of his informal concept of addition as an incrementing process and thus applied a counting-on procedure ("66; 67, 68, 69, 70"). He interpreted the score-difference problem (30 points beats 8 by how many) in terms of a how-many-more concept and so used a counting-up procedure (start with 8 and count how many units it takes to get up to 30). His informal-solution procedures often relied on concrete supports, such as making marks to keep track of score differences. Indeed, the use of concrete supports was, for many problems, necessary for success.

PROBLEM MODELING. Research (e.g., Carpenter et al., 1981; Carpenter & Moser, 1982; Gibb, 1956) indicates that young children also use informal strategies that model the meaning of basic addition and subtraction problems. Indeed, if children are introduced to a problem with which they are not familiar, they tend to construct a strategy by directly modeling the meaning of the problem. Initially, children use real objects to represent the quantities described in a problem, perform actions to imitate the arithmetic operation indicated in the problem, and count to determine the answer (Carpenter, in press).

For *change* problems ($A \leftarrow B = \_$), which imply something is added to an existing set to make it larger, and *combine* problems (A+B=__), which imply that two (sub)sets are joined (to form a whole), 5- to 6-year-olds typically use a concrete counting-all strategy. For the change or combine problem described in Table 14.1, this would entail counting out two objects to represent two marbles, producing four objects to represent four marbles, and then counting all the objects put out.

For *take-away* problems (C − A = __), which imply that something is removed from an existing set, children resort to separating from. For the take-away problem in Table 14.1, the child would construct a set of six objects to represent the original quantity (six marbles), count and remove two marbles to model the taking-away action, and then count the remaining objects to determine the answer: four.

For *missing-addend* problems (A + __ = C), in which only the first subset and the whole are given; *additive subtraction* or *equalize* problems (A← __ = C), in which an initial amount is given and an amount that must be *added* to produce a given total must be determined; and *missing-augend* problems (__←B = C), in which the amount added and the total are given, 5- to 6-year-olds usually use a different strategy: adding on. To solve the missing-addend problem in Table 14.1, a child would count out two objects to represent the initial quantity (two marbles), add more objects until there were six, and finally count the added objects to determine the answer: four.

For *compare* problems (A↔B = C), which imply finding a difference between two quantities, 5- to 6-year-olds most frequently use a very different strategy from those described above: matching. To solve the compare problem in Table 14.1, a child would put out two objects to represent the first quantity (two red marbles), count out four objects to represent the second quantity (four red marbles), line the two sets up (in one-to-one correspondence), and count the extra objects of the second set to determine the difference: two.

As they develop, children use mental strategies rather than concrete strategies to solve word problems. Initially, the child's choice of mental strategy also reflects or models the type of word problem presented. As Table 14.1 shows, simple addition (change and combine) problems are solved by CAF or COF. Take-away problems are solved by counting down, whereas missing-addend, additive-subtraction, missing-augend, and compare (or difference) problems are solved by counting up. Thus both concrete and mental strategies can directly model the meaning of the word problems.

Because change problems ("A and B *more* equal") more directly model their understanding of addition as an incrementing process than do combine problems ("A and B together equal"), change problems may initially

TABLE 14.1: The Concrete and Mental Counting Strategies That Model Different Types of Addition and Subtraction Word Problems

Problem Type	Example	Solution Procedure: Concrete Strategy	Solution Procedure: Mental Strategy
Change (A ← B = __)	Al has two marbles, and he buys four more. How many does he have all together now?	Concrete counting all	CAF or COF
Combine (A + B = __)	Al has two red marbles and four blue marbles. How many does he have all together?	Concrete counting all	CAF or COF
Take away (C − A = __)	Al had six marbles. He lost two in a match. How many marbles does Al have left?	Separating from	Counting down
Missing addend (A + __ = C)	Al has six marbles all together. Two are red, and the rest are blue. How many blue marbles does Al have?	Adding on	Counting up
Additive Subtraction (A ← __ = C)	Al has two marbles. He needs six marbles to play in a match. How many more marbles does he need to buy at the store?	Adding on	Counting up
Missing augend (__ ← B = C)	Al had some marbles. He bought four more at the store. Now he has six. How many did he have to start with?	Adding on	Counting up
Compare (A ↔ B = C)	Al has two marbles, and Bert has four marbles. How many more marbles does Bert have than Al?	Matching	Counting up

be more meaningful to children. Typically, though, children quickly assimilate combine problems to informal arithmetic knowledge, and they soon solve change and combine problems with equal facility (Baroody & Ginsburg, in press). Because simple addition (change and combine) problems and take-away problems are more easily modeled than other types of problems, children successfully solve these more direct problems earlier. Missing-addend problems ($A + \_\_ = C$) represent a different level of difficulty. Missing augend problems ($\_\_ \leftarrow A = C$) are very difficult because the initial quantity is unknown and, as a consequence, the problem cannot be readily modeled (Carpenter, in press).

CONCRETE AIDS. Because children's first informal procedures require concrete objects (e.g., fingers, blocks, marks), early problem-solving success will depend upon whether or not concrete aids are available or allowed. In fact, research (e.g., Lindvall & Ibarra, 1979; Riley, Greeno, & Heller, 1982) indicates that the availability of concrete objects is—at least at first—very important to young children's arithmetic problem-solving success. For example, Hebbeler (1977) gave 4- and 5-year-old preschool children addition and subtraction word problems like: "The elephant and the owl went shopping. The elephant brought along five pennies, and the owl brought along seven. How many pennies did they bring altogether?" In contrast to children who were not shown objects and who did poorly, most children who were given concrete representations of the problem were quite successful. It appears that most children just entering school can solve simple addition problems when concrete objects are available but appear far less competent without concrete aids (Carpenter & Moser, 1984).

Strategy Flexibility

At first, children do not solve word problems flexibly. Below the middle of first grade, most children can only represent and solve problems by directly modeling the actions or relationships in the problem. For example, it appears that early on children do not recognize that subtraction can be thought of in different ways: as take-away, additive-subtraction, and compare (difference) problems. Thus initially they may not recognize the interchangeability of informal subtraction strategies and hence use a particular strategy only with a particular type of word problem: separating from or counting down with take-away problems, adding on or counting up for missing addend problems, and matching or counting up for compare or difference problems.

This helps to explain why Adam always computed score differences by counting up and consistently rejected using his written subtraction algorithm. Adam's conception of difference problems was not connected to his representation of "take-away." Because he interpreted written subtraction as "take-away," he did not see a connection between his formal subtraction algorithm and difference problems. In other words, because he did not represent difference problems as subtraction, he failed to see that the formal subtraction algorithm was applicable.

Children may become more flexible in their choice of strategy because their knowledge becomes more interconnected. For example, once children view "take-away," missing-addends, and compare (difference) as alternate conceptions of subtraction, they may see that the strategies that were tied to these conceptions are interchangeable. Hence, when presented with a difference problem, say, the child might use counting down or counting up, whichever required less effort.

Individual Differences

Some research (e.g., Lindvall & Ibarra, 1979) indicates that word problem-solving ability, even simple arithmetic problems with concrete objects available, cannot be taken for granted in all preschoolers—especially disadvantaged children. Carpenter & Moser (1984) found that about one-seventh of their entering first graders were unable to solve any addition word problem even by concrete counting all. Ginsburg & Russell (1981) found that lower-class children had many informal mathematical strengths. However, an inability to solve simple arithmetic problems stood out as a deficiency. Furthermore, children with learning handicaps may have great difficulty in getting beyond the level of direct modeling.

PROBLEM SOLVING AND WORD PROBLEMS

Problem-Solving Deficiencies

A key objective and a major focus of mathematics instruction should be the development of problem-solving ability (National Council of Supervisors of Mathematics, 1977; National Council of Teachers of Mathematics, 1980). However, the National Assessment of Educational Progress (NAEP)(1983) suggests that American schools are relatively successful in teaching skills but less successful in encouraging competencies that require understanding (conceptual learning and problem solving). More specifically, the assessment indicates that most children master basic number skills, such as ordering numbers and matching written words with symbols (e.g., matching "sixty-seven" and 67"); basic number combinations, such as $6 + 8 = 14$ and $5 \times 7 = 35$; and whole-number computational skills, including renaming algorithms. Children are also generally successful in solving routine one-step word problems, such as those found in typical textbooks (Carpenter, Matthews, Lindquist, & Silver, 1984). However, national assessments show that the majority of students at all ages have difficulty with nonroutine problems that require some analysis or thinking (e.g., Carpenter, Corbitt, Kepner, Linquist, & Reys, 1980; Carpenter et al., 1984).

For example, when extra information was included in problems, some children just mechanically applied an arithmetic operation to all the numbers given. For the problem below, a common answer is "nine," because all three numbers mentioned are added together $(2 + 3 + 4)$. Many children apparently do not analyze the problem to determine what information is needed and what information is not needed.

> Jim bought a bag of marbles. There were 2 red marbles and 3 blue marbles in the bag. The bag of marbles cost 4 cents. How many marbles did Jim buy altogether?

Why the discrepancy in performance between routine and nonroutine problems? Primary-level children are frequently successful on routine problems because simple arithmetic problems are meaningful and have a straightforward format. Routine problems can be readily assimilated to children's informal arithmetic knowledge. Moreover, the format of routine problems does not demand much analysis. Consider the simple change and combine addition problems in Table 14.1. The basic task required to solve these problems is identifying which operation is appropriate (LeBlanc, Proudfit, & Putt, 1980). This can be done rather readily by just a superficial reading of problems. In contrast, nonroutine problems do not automatically connect with children's existing knowledge of concepts and procedures. Furthermore, the format of nonroutine problems requires more than a simple-minded identification and application of known arithmetic operation.

Requirements for Effective Problem Solving

Problem solving with nonroutine problems requires a thoughtful analysis: defining the problem, planning a solution strategy, implementing the solution strategy, and checking the results (Polya, 1973). A thoughtful analysis entails understanding, problem-solving skill, and drive. Effective problem solving also requires flexibility.

UNDERSTANDING. The first step in understanding a problem is to define clearly the nature of the problem: What is the unknown or goal of the problem? (Carpenter et al., 1984). This helps in deciding what information is needed to solve the problem (and what information is irrelevant), what solution methods are appropriate (and which are inappropriate), and what solutions are reasonable (and which indicate the need for further effort). Children who do not clearly identify the goal of the problem may well have difficulty choosing and applying a solution procedure and checking their results. For example, when Adam was asked by how many his score of 30 beat a score of 8, a clear understanding of the goal of the problem might have helped him to realize that his strategy of writing down the numerals from 8 to 30 was not sufficient and that his response of "30" was not possible.

Defining the unknown is especially important with nonroutine problems. Consider the following exchange reported by Lindvall and Ibarra (1980):

> I: Chuck had five toys.
> G: Put out five [blocks].
> I: Chuck then found some more toys.
> G: Put out three.
> I: Then he had nine toys altogether.
> G: Count all of them. [Counts the eight blocks.] Put out one more.
> . . .
> I: How many toys did Chuck find?
> G: . . . [counts the nine blocks].

The child did not correctly identify the goal of the additive-subtraction problem as finding the amount that must be added to five to make nine. As a result, G. chose an incorrect solution procedure (counting all) and was satisfied with the impossible answer nine.

Understanding a problem implies having an appropriate mental representation, which involves having a sufficient body of facts and concepts (Riley et al., 1983). Without adequate knowledge to understand (mentally represent) the problem, the child has very little basis for choosing and implementing a solution strategy and for critically checking the results. For instance, in the case reported above, G. may not have had a conceptual basis for assimilating additive-subtraction problems. As a result, the child simply interpreted the problem in terms of a familiar notion (as a change problem). Thus the child chose a solution procedure and gave answers that were appropriate for the interpretation but inappropriate for the problem. Difficulties with problem representation increase as children advance in school and are expected to acquire a broader and more complex range of mathematical knowledge (Silver & Thompson, 1984).

PROBLEM-SOLVING SKILLS. When faced with nonroutine problems—in which the unknown, solution procedure, and answer are not obvious—it may be helpful to use certain problem-solving aids. Skills or devices that aid in the analysis of a problem are called heuristics (Polya, 1973). One heuristic for better analyzing a problem is to make a drawing that represents the problem (e.g., see Davis & McKillip, 1980; LeBlanc et al., 1980). A drawing can help a child to define the problem and decide on a solution procedure (operation) (see Figure 14.1). If children do not have skills for systematically analyzing and evaluating problems, it will be more difficult for them to solve nonroutine problems.

DRIVE. Children must have more than a capacity to understand problems and the skills to analyze new problems; they must have the drive to put forth the effort required for a thoughtful analysis. This drive comes from interest, self-confidence, and perseverance. Children willingly lavish energy on activities that interest them, and they are stingy with activities that appear irrelevant or unimportant to them. Moreover, problem solving

FIGURE 14.1: Draw-a-Picture Heuristic

Pictures can help children to define the problem (steps 1 and 2), plan and carry out a solution procedure (steps 3 and 4), and check the answer (reject impossible answers).

(1) Joe has 2 white marbles and (2) some blue marbles.

(3) Altogether he has 6 marbles.

(4) Of the 6 marbles, how many are blue? (5) Two, four, six, or eight?

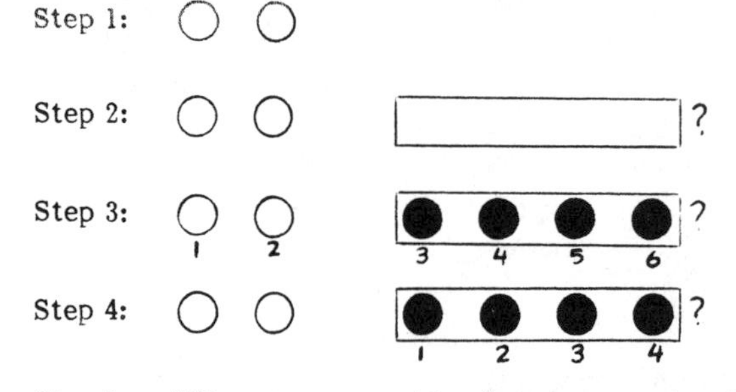

Step 5: "The answer must be four because two [white] four [blue] are six all together and that's what he had." (So the answer could not be two [the difference of four and two], six [the total], or eight [the sum of two and six]).

entails making choices, coping with uncertainty, and accepting the possibility of failure. In brief, problem solving entails risk (House, 1980) and hence requires the confidence to take risks. Finally, problem solving requires time to think and explore. It may involve making mistakes, uncovering errors, or beginning again. Thus problem solving requires a commitment to stay with a problem. In summary, the affective factor of drive, which draws upon interest, self-confidence, and perseverance, is important to nonroutine problem-solving success (Lester, 1980; Silver & Thompson, 1984).

FLEXIBILITY. At the heart of problem solving is flexibility: the ready adaptation of existing resources to meet the demands of a new task. Flexibility is promoted by a combination of understanding (a rich, well-integrated body of knowledge), problem-solving skills (heuristics), and drive (interest and self-confidence). A rich, well-integrated body of knowledge permits a child to more readily make sense of new tasks and see new connections. Heuristics, such as "draw a picture," may help the child to look at a problem in different ways and try a wider range of solution procedures. When interested in pursuing the problem, a child is far more likely to apply understanding and problem-solving skills to a new task. Furthermore, a child who is willing to engage in problem solving without feeling threatened by the prospect of ridicule or failure is freer to engage in a range of ideas. In brief, the development of flexibility (see Example 14.1) and problem-solving ability are complex processes that involve nurturing meaningful learning, problem-solving skills, and drive.

Traditional Problem-Solving Instruction

Often problem-solving instruction at the primary level involves little more than assigning textbook word problems. Frequently word problems are assigned *after* an operation has been introduced with the aim of (a) mastering the basic number facts of the operation, (b) practicing the computational algorithm(s) related to the operation, and (c) strengthening specific, real-world applications of the operation (LeBlanc et al., 1980). Thus the aim of word-problem exercises is the practice of basic skills, *not* the development of problem-solving ability.

Because the aim is to practice basic skills, "problem-solving" exercises often consist of a series of routine word problems. Such word problems provide only the specific information needed to perform the intended operation. The correct solution procedure is obvious: the operation to be practiced. Often all the problems in an exercise require the same operation.

EXAMPLE 14.1

The Three Utilities Problem: Encouraging Flexibility

One way to encourage flexibility is to give children problems that require them to question an assumption or to break with a habit. One interesting example of such a problem is the three utilities problem.

> The problem: A housing developer built three houses. He had to connect each house to three utilities: electric, gas, and water. However, the electric, gas, and water lines could not cross each other. Draw in the connections without crossing any of the utility lines.

Seems like it should be an easy problem, right? Try it and see.

Give up? The key to solving the problem is avoiding or overcoming an assumption. Many people run into difficulty because they assume that the utility lines cannot cross the lines representing the houses. Once free from the unnecessary restriction, the solution is easy: Just pass a utility line under the houses.

Every problem encountered has a correct and only one correct answer. Finally, because the aim is to automatize fact computational and application skills, pupils are expected to complete exercises quickly.

Routine word-problem exercises are unlike real-life or mathematical problem solving. In Table 14.2, the characteristics of routine word problems are compared with those of genuine problems. With routine word problems, even a superficial reading will reveal the unknown, the data needed, and the solution procedure (arithmetic operation needed). In contrast, genuine problems require a thoughtful analysis to define the unknown, the data needed, and a solution strategy. The unknown of real problems may not be clearly specified and an analysis may be required to spell out exactly the goal of the problem. Moreover, real problems often contain a great deal of information. Quite often the problem solver must sift through information, discard unneeded data, and locate data necessary to solve the problem. Sometimes, even with a huge amount of information available, the data needed to solve the problem may not appear. Thus the problem may not be solvable unless the needed data can be collected. Because genuine problems can be solved in a number of ways, thought is needed to choose the most efficient or elegant solution strategy. Because there are times when solution procedures are not at all obvious, thought is needed just to find *any* solution procedure. In addition, unlike routine word problems in which there is one correct solution, real problems may have numerous answers or no answers at all. Finally, very much unlike routine problems, the thoughtful analysis required by nonroutine problems is usually time-consuming. Clearly exercises consisting of routine problems, especially those in which the same operation is applicable to every problem, are uncharacteristic of genuine problem solving and do little to advance problem-solving ability.

Even when an attempt is made to teach problem-solving skills, the focus is often quite narrow. Children are frequently taught heuristics that have limited utility and may even be misleading. A heuristic commonly emphasized to children is the "keyword" approach: Look for the word that signals

TABLE 14.2: A Comparison of Routine Word Problems and Genuine Problem Solving

Routine Word Problems Commonly Found in School Texts	Problem-Solving Endeavors Common to Everyday Life and Mathematics
The unknown is specified or readily apparent.	The unknown may not be specified or obvious.
Only the specific information needed to calculate the answer is provided.	Too much (or too little) information is available.
One correct solution procedure is obvious.	Many solution procedures, which may or may not be apparent, apply.
There is <u>one</u> correct solution.	There may be a number of answers or even no answer.
The solution should be arrived at quickly.	Significant problems are often solved slowly.

an operation. For example, "Whenever you see 'and,' it means add the numbers." Or, "Whenever you see 'left,' it means subtract the numbers." Such tricks can be quite successful with exercises involving routine problems. For example, 97% of the problems in one commonly used textbook could be solved using this superficial approach (Schoenfeld, 1982).

However, a keyword approach is less helpful when children encounter nonroutine problems (and many real-world problems). Rote or hasty application of keyword rules may actually lead to bizarre answers. For instance, given a compare problem (e.g., "Lee has 5 dolls, *and* Donna has 2 dolls. How many more dolls does Lee have than Donna?"), a child who is mechanically following the "and-means-add" rule will answer, "Seven." A child who blindly follows a left-means-take-away rule may subtract with a problem such as, "When Mrs. Jones' class went out for recess, the boys left 5 jackets and the girls left 3. How many jackets were left?" (Kilpatrick, 1985). Thus superficial tricks for solving word problems may work with a limited range of problems but may actually interfere with broader problem-solving efforts—with the child's natural and often powerful ability to analyze problems (Carpenter, in press; Carpenter et al., 1980).

In addition, the nature of word-problem exercises often undermines drive. Problems are often unrelated to children's interest. Because they are essentially repetitive practice, such exercises are frequently unchallenging and boring (Kline, 1974). Moreover, concrete aids are sometimes not allowed, and this undermines some children's confidence. When nonroutine problems are introduced, some children are intimidated because they realize that they cannot solve the problem quickly. In brief, children often have little reason to approach word-problem exercises with excitement, and sometimes they have reason to be afraid.

EDUCATIONAL IMPLICATIONS: INSTRUCTION THAT ENCOURAGES PROBLEM SOLVING

How can educators exploit word problems to make mathematics instruction more meaningful to children? How can educators nurture problem-solving competence—that is, analytic skills, drive, and flexibility? Delineated below are general guidelines and examples.

INTEGRATING WORD PROBLEMS WITH INITIAL INSTRUCTION. A direct implication of recent research is that the introduction of word problems need not be deferred until basic formal skills have been mastered but should be integrated into the primary math curriculum from the start (Carpenter & Moser, 1984). Routine addition and subtraction problems can, in most cases, be introduced to children just beginning formal mathematics instruction. Indeed, word problems can be an important bridge between a child's informal concepts and procedures and school-taught symbolic mathematics.

1. *Introduce operations with word problems.* Because word problems can be more meaningful to children than formal mathematics, it may be helpful to introduce the definitions and symbols of operations with word problems (e.g., see Carpenter et al., 1980; Quintero, 1985). For example, to introduce the formal symbolisms of addition, expressions that include the plus sign can be linked to simple word problems. Because they may more readily connect with children's informal understanding of addition as an incrementing process, use change problems first and then both change and combine problems. This may be especially important to do with children with learning difficulties.

To introduce the concept and written representation of multiplication, children might be asked to analyze word problems like the following:

> There are 8 pieces of gum in a package. Jimmy buys 4 packages. How many pieces of gum does he have?

By, say, drawing a picture of four rows of eight sticks, the child can solve the problem by counting. Some children may (rather quickly) discover that the problem can be solved by repeated addition: $8+8+8+8$. At this point, children can be helped to see that the repeated addition of like terms can be represented as 4×8—as four groups (packs) of eight things (pieces of gum). By relating the formal symbolism and definition to real problems and objects, the formal mathematics should make more sense.

2. *Relate formal representations to word problems.* Word problems can help to make abstract formal representations concrete and understandable. For example, children have some trouble making sense of atypical equations such as $4 + \_ = 6$ and $\_ - 2 = 6$ even though they can solve missing-addend problems and missing-minuend problems by counting up. By helping children to see that the format $A + \_ = C$ represents a missing-addend problem and the format $\_ - B = C$ represents a missing-minuend problem, the formal symbolism can be connected to a child's informal understanding and solution procedures and, hence, take on meaning (Carpenter, in press; Carpenter & Bebout, 1985; Bebout, 1986).

3. *Encourage the use of concrete aids to solve problems.* Especially at first or when using word problems to introduce formal mathematics, concrete aids should be available and their use encouraged.

USING NONROUTINE WORD PROBLEMS. Though more nonroutine word problems are beginning to appear in textbooks (LeBlanc et al., 1980), greater emphasis should be placed on such problems. The use of nonroutine problems is more likely to encourage thoughtful problem solving and a more realistic view of genuine problem solving.

1. *Use problems that require an analysis of the unknown.* Rather than give children problems that simply entail computing an answer, use problems that require them to use the computed results in a thoughtful manner. For example, transform routine problems (e.g., Jim has $2 and Karl has $3. How many dollars do they have altogether?) into nonroutine problems (e.g., Jim has $2 and Karl has $3. Do they have enough to buy a baseball glove that costs $6?).

2. *Use problems with too much, too little, or incorrect data.* This requires thoughtful analysis of the unknown and the data. To introduce the issue of analyzing the data for needed and unneeded information, it may be helpful to include just one extra bit of information, as in the following problem.

> Jill needs 4 candles for her birthday cake; Kate needs 6. There are 12 candles in a box. How many candles do the girls need altogether?

Once children are accustomed to looking for extra information, more complicated problems, like that below, can be assigned.

> Mrs. King's first-grade class won the most medals during field day: Amy won 2 blue ribbons; Bill won 1 blue and 1 red; Chuck won 2 red; Diane won 2 blue and 1 red; Ed won 3 red; George won 3 blue, and Harriet won 2 blue and 2 red. How many red ribbons did the boys in Mrs. King's class win?

Children should regularly be exposed to problems with insufficient data (e.g., Problems A, B, and C below) or incorrect data (e.g., Problem D below).

> A. Jack had 6 marbles. He bought some more. How many marbles does he have now?
>
> B. Sarah's books cost $7 and her pencils $2. How much change did she receive?
>
> C. Ginny had 9 candies. She gave some to her sister Beth and some to her brother Eric. Ginny had 5 left that she kept for herself. How many candies did Ginny give her sister Beth?
>
> D. Sam bought 6 books and 4 pencils. How much did he have to pay the storekeeper altogether?

3. *Use problems that can be solved in more than one way.* Children regularly should be given exercises in which there is more than one possible

solution method. This will help them appreciate that most problems can be solved in a number of ways. Consider the example below:

> Toni is in the candy store. She has 9 pennies. She wants a jawbreaker that costs 5 cents, but also a piece of bubble gum that costs 3 cents. Can she buy both of them?
>
> *Procedure 1*: $5 + 3 = 8$. The total cost of 8 cents is less than she has (9 cents), so she can buy both.
>
> *Procedure 2*: $9 - 5 = 4$, $4 - 3 = 1$. After taking away the cost of the jawbreaker (5 cents), and then the cost of the bubble gum (3 cents), she still has one cent left. Therefore, she can buy both.

4. *Use multistep problems.* Routine problems involve one step: the single application of one operation. Multistep problems involve several applications of an operation or more than one operation. Consider the following problem:

> Sandra had 95 cents. She spent 25 cents on gum and 50 cents on candy. How much money did Sandra have left after buying the gum and candy?

This two-step problem could either be solved by two applications of the subtraction operation ($95 - 25 = 70$; $70 - 50 = 20$) or by applying two operations ($25 + 50 = 75$; $95 - 75 = 10$). Such multistep problems require thoughtful analysis of the unknown, the data, and the solution method.

5. *Use problems with more than one possible answer.* Children should become accustomed to the fact that some problems may have more than one possible answer.

> Toni is in the candy store. She has 20 pennies. Jawbreakers cost 5 cents and bubble gum costs 3 cents. What can Toni buy?
>
> *Answer 1*: 4 jawbreakers ($5 + 5 + 5 + 5 = 20$).
>
> *Answer 2*: 3 jawbreakers and 1 piece of gum ($5 + 5 + 5 + 3 = 18$).
>
> *Answer 3*: 2 jawbreakers and 3 pieces of gum ($5 + 5 + 3 + 3 + 3 = 19$).
>
> *Answer 4*: 1 jawbreaker and 5 pieces of gum ($5 + 3 + 3 + 3 + 3 + 3 = 20$).
>
> *Answer 5*: 6 pieces of gum ($3 + 3 + 3 + 3 + 3 + 3 = 18$).

6. *Include problems that require an extended effort.* To get used to the idea that some problems require perseverance, children need exposure to problems that cannot be solved immediately. One way to accomplish this end is to create problems of personal interest in which the data must be gathered over time. For example, how many runs will you get altogether next week during the daily kickball game?

MIXING PROBLEMS. To encourage children to analyze word problems thoughtfully, a variety of problems should be used. Mixing the problems requires the child to define the unknown and suitable representation of the problem. Incidentally, it also rewards flexibility.

1. *Mix the operations required.* Especially with routine problems, word-problem exercises should involve more than one operation. A mixture of operations will encourage children, even with routine problems, to analyze more carefully the unknown, the information given, and the solution procedure needed.
2. *Mix the types of problems.* Instead of relying extensively or exclusively on simple addition and subtraction word problems, mix change, combine, and take-away problems with missing-addend problems, compare problems, and so forth. Mixing problem types demands a more careful analysis of the unknown, the data given, and the solution method.

TEACHING AND ENCOURAGING HEURISTICS. Even primary-level children can be taught basic heuristics that may help them to analyze problems more thoughtfully and flexibly. Drawing a picture has already been mentioned as one way that may help to define and understand the problem.

A heuristic for planning a solution strategy is determining if the problem resembles a familiar problem. For example, relatively difficult missing-augend problems resemble somewhat more familiar missing-addend problems in that, for both, one part and the whole is given and one part is missing.

A heuristic for checking the results is to just make an intuitive estimate of the answer and then check the solution against the estimate. A solution that does not make sense when compared to the estimate signals that something is amiss. Consider the following missing-addend problem.

> Roberto had 3 pencils. He bought some more. He now has 10 pencils. How many pencils did Roberto buy?

Even a global estimate could be helpful. If the child estimated that the answer had to be less than 10 but used a counting-on procedure to arrive at the answer of 13, the child might realize that his or her solution procedure did not fit the problem and should be abandoned.

SUMMARY

Even before they receive formal arithmetic instruction and have mastered written addition and subtraction, children can solve simple arithmetic word problems because they rely on concrete objects and use solution strategies that directly model their understanding of addition and subtraction. On the whole, U. S. schoolchildren do well on routine word problems but not on nonroutine word problems that require thoughtful analysis. Solving nonroutine problems requires understanding, use of heuristics, drive, and flexibility. Routine word problems do not make such demands on children because the unknown, data, and solution procedure are readily apparent. Routine problems are unlike real problems in that there is always *a* correct answer that can be determined quickly. Problem-solving competencies can be encouraged by integrating word problems with initial instruction, regular use of nonroutine problems, mixing problems, and encouraging the thoughtful application of heuristics.

CHAPTER 15

Epilogue

PROGRESS IS POSSIBLE: THE CASE OF ADAM

When I first met Adam, pain and hopelessness seemed etched in the learning-disabled lad's every move. He was stiff, reticent, passive, and unhappy. He was confused and unsure of himself. He had little interest in mathematics or much of anything. Indeed, he even had difficulty learning the procedures of math games that I introduced and played the games without enthusiasm. All in all, Adam was a sad child to behold.

Adam learned much from an informal approach. There are basic skills and concepts that he must yet master, and there are aspects of basic mathematics that perhaps he will never learn. Nevertheless, Adam has begun to master basic number combinations, computational algorithms, place-value notions, and mental arithmetic with multidigit terms. Perhaps more important, Adam now has more spark and seems more capable of enjoying himself. He is willing to give others a chance to teach him and to give himself a chance to learn. In brief, he seems more comfortable with himself, others, and mathematics. For example, in Session 28 (3/10/82), Adam was asked to play an estimation game, which involved the addition of two-digit addends and a time limit of two seconds. In contrast to his initial reluctance to engage in estimation, he responded with enthusiasm: "Let her rip!"

Adam is more willing to assert himself, less defensive, and more confident in his abilities. In Session 22 (1/13/82), for example, Adam explained that adding across (horizontally) was more difficult than regular addition (vertically).

INTERVIEWER: Why is it harder to do it this way [horizontally] than this way [vertically]?

ADAM: It's um, it's like harder to carry and you get mixed up. And I don't know how to carry it so good.

INTERVIEWER writes:

$$\begin{array}{r} 38 \\ +17 \\ \hline \end{array}$$

ADAM: Oh, that's *easy*. [Adam proceeds to correctly use the carrying algorithm.]

In short, Adam demonstrated a willingness to discuss a problem that was bothering him, accurately described his difficulty, and confidently employed a now-familiar procedure.

TOWARD MEANINGFUL AND ENJOYABLE MATHEMATICAL LEARNING

Cognitive theory suggests that mathematical competence builds slowly from the concrete and specific to the abstract and general. This fundamental principle of development applies to mathematical learning of all kinds at all levels: whether a child is learning to count, add, write multidigit numerals, execute a borrowing algorithm, or solve arithmetic word problems. Though children can increasingly profit from verbal explanations, demonstrations, and symbolic textbook presentations, early skills and concepts emerge from concrete activities. Indeed, even adolescents and adults learn new topics in mathematics best when instruction is introduced concretely and then proceeds to the abstract.

Like almost anyone's initial knowledge of almost any topic, children's first understanding of a mathematical subject is likely to be intuitive. This intuitive (unanalyzed or impressionistic) knowledge is typically quite context bound and unsystematic (Lunkenbein, 1985). For example, by observing the effects of adding objects to small sets, very young children conclude that addition "makes more." However, this intuitive grasp of addition leads a child to believe that when a candy is added to a cup with seven candies, that cup then contains more than a cup of nine to which nothing has been added.

In time, learners formulate more precise relationships in a domain. Because this is a gradual process, a learner does not achieve a coherent system at once. Early knowledge may be of limited generality and logically inconsistent. For example, kindergarteners realize that counting is quite useful for defining sums precisely. They can figure out that seven candies and one more are eight, and that this is less than nine candies. However, young children may not appreciate commutativity—that, say, $7 + 3$ and $3 + 7$ are equivalent (*both* have a sum of 10). Though $7 + 3 = \_$ and

3 + 7 = __ are viewed as different problems, a young child may nevertheless determine the sums of *both* problems by using the same procedure (e.g., by counting all starting with the larger term: "1, 2, 3, 4, 5, 6, 7; 8 is one more, 9 is two more, 10 is three more—10"). By disregarding addend order to compute the sums, the child, in effect, treats 7 + 3 = __ and 3 + 7 = __ as the same problem. Though it seems illogical to adults, children may add numbers in either order because they believe that they will get the *correct answers* though not necessarily the *same answer* (Baroody & Gannon, 1984; Baroody & Ginsburg, in press).

With development, children learn more relationships and organize their knowledge more coherently. As their knowledge forms a more complete and logical system, children are better able to see logical implications and reason deductively (applying general principles to solve specific problems). For instance, shown the problems 700 + 100 = 800 and 100 + 700 = __, a school-age child immediately recognizes that the problem pairs will have the same sum because the general principle of commutativity applies. Thus the child can quickly deduce that the sum of 100 + 700 is 800.

Young children seem to have a natural interest in informal mathematics. Like Alison, described in Chapter 2, or Aaron in Chapter 8, preschoolers spend hours practicing and applying counting, number, and arithmetic skills. Because these activities are an important part of their lives and because they have numerous opportunities to use and reflect upon these activities, children learn much about these informal domains. Knowledge of counting, number, and arithmetic gradually increases and becomes more interconnected. Indeed, the typical child entering kindergarten understands a range of counting principles that constitutes a relatively abstract grasp of counting and number. Such children can construct concrete and mental computing strategies and solve simple arithmetic problems, which indicates a significant understanding of addition and subtraction. In brief, children come to school with an active mathematical intelligence and curiosity.

Once in school, children all too frequently become passive learners and use mathematics mindlessly. Most lose interest in mathematics. Some children, like Adam, are devastated intellectually and emotionally. Children learn to learn mathematics passively and do mathematical exercises mechanically. They learn to be helpless because it seems that no matter what they do, failure is inevitable (Reyes, 1984). In too many cases, it is the nature of formal instruction that dampens children's interest and enthusiasm and creates fear and apathy.

School mathematics is too often meaningless and joyless to children, if not confusing and threatening. In contrast to informal learning experiences, instruction based on absorption theory does not give children the means, incentive, or time to develop a meaningful and coherent system of mathematical knowledge. Such programs regard mathematics as primarily a matter of writing symbols on paper according to prescribed rules (Davis, in press). Such instruction overlooks that children need a "reality" to write about and that everyday mathematics is less dependent on using written symbols than on understanding (Davis, in press). Instead of an ideal opportunity to encourage thoughtful analysis, pattern searching, and numerical reasoning (i.e., flexible problem solving), the basic number combinations are treated as a basket of facts that must be mastered quickly through rote memorization. Instead of helping children to view computation as a means to more important ends, efficient computational skill becomes an end in itself. Instead of occasions for encouraging thinking ability, estimation and word-problem exercises (when used at all) are used to reinforce rotely learned skills.

Instruction that disregards children's informal knowledge and too quickly introduces formal symbols and rules *forces* children to learn rotely rather than meaningfully. Mechanical behavior, bizarre answers, and learning problems frequently stem from forcing children to rotely memorize mathematics, not from a lack of natural talent or drive. Rotely memorized mathematics typically remains unconnected to more practical knowledge or other formal mathematics. When children are required to push around symbols that have no meaning, they often produce and are untroubled by bizarre answers. Furthermore, some children, like Mark described in Chapter 5, simply refuse to play along. They give up and fail to memorize formal mathematics altogether.

Mathematics need not be a meaningless and unpleasant chore that children are forced to endure. Basic mathematics instruction can and should be meaningful and enjoyable for children at all levels of ability. To achieve this, teachers should build on children's informal mathematics, patiently wait for readiness and insight, and provide interesting opportunities to learn and practice mathematics. The rewards are well worth the effort and wait: children actively and eagerly involved in mathematical learning and thinking—children who enjoy mathematics, school, and themselves as learners. By building carefully on children's practical mathematics, teachers can nurture the mathematical intelligence and curiosity that children bring with them to school and help children of all levels of ability to grow and achieve their potential.

References

Acredolo, C. (1982). Conservation–nonconservation: Alternative explanations. In C. J. Brainerd (Ed.), *Children's logical and mathematical cognition* (pp. 1–31). New York: Springer-Verlag.

Allardice, B. S. (1978, July). *A cognitive approach to children's mathematical learning: Theory and applications.* Paper presented at the American Association of School Administrators, Minneapolis.

Allardice, B. S., & Ginsburg, H. P. (1983). Children's learning problems in mathematics. In H. P. Ginsburg (Ed.), *The development of mathematics thinking* (pp. 319–349). New York: Academic Press.

Almy, M. (1971). Longitudinal studies related to the classroom. In M. F. Rosskopf, L. P. Steffe, & S. Taback (Eds.), *Piagetian cognitive development research and mathematical education* (pp. 215–241). Washington, DC: National Council of Teachers of Mathematics.

Anderson, R. C. (1984). Some reflections on the acquisition of knowledge. *Educational Researcher*, *13*(9), 5–10.

Apostel, L., Mays, W., Morf, A., & Piaget, J. (1957). Les liaisons analytiques et synthétiques dans les comportements du sujet. *Etudes d'épistémologie génétique.* Vol. 4. Paris: Presses Univer. France.

Ashcraft, M. H. (1982). The development of mental arithmetic: A chronometric approach. *Developmental Review*, *2*, 213–236.

Ashcraft, M. H. (1985, April). *Children's mental arithmetic: Toward a model of retrieval and problem solving.* Paper presented at the biennial meeting of the Society for Research in Child Development, Toronto.

Baroody, A. J. (1983a, April). *The case of Adam: A specific evaluation of a math learning disability.* Paper presented at the meeting of the American Educational Research Association, Montreal.

Baroody, A. J. (1983b). The development of procedural knowledge: An alternative explanation for chronometric trends of mental arithmetic. *Developmental Review*, *3*, 225–230.

Baroody, A. J. (1984a). The case of Felicia: A young child's strategies for reducing memory demands during mental addition. *Cognition and Instruction*, *1*, 109–116.

Baroody, A. J. (1984b). Children's difficulties in subtraction: Some causes and cures. *Arithmetic Teacher 32*(3), 14–19.

Baroody, A. J. (1984c). Children's difficulties in subtraction: Some causes and questions. *Journal for Research in Mathematics Education*, *15*, 203–213.

Baroody, A. J. (1984d). More precisely defining and measuring the order-irrelevance principle. *Journal of Experimental Child Psychology*, *38*, 33–41.

Baroody, A. J. (1984e). A re-examination of mental arithmetic models and data: A reply to Ashcraft. *Developmental Review*, *4*, 148–156.

Baroody, A. J. (1985a). Mastery of the basic number combinations: Internalization of relationships or facts? *Journal for Research in Mathematics Education*, *16*, 83–98.

Baroody, A. J. (1985b, April). *Mental addition protostrategies: Retrieval or problem solving?* Paper presented at the biennial meeting of the Society for Research in Child Development, Toronto.

Baroody, A. J. (1986). The value of informal approaches to mathematics instruction and remediation. *Arithmetic Teacher*, *33*(5), 14–18.

Baroody, A. J. (1989). *A guide to teaching mathematics in the primary grades*. Boston: Allyn and Bacon.

Baroody, A. J. (in press). The development of counting strategies for single-digit addition. *Journal for Research in Mathematics Education*.

Baroody, A. J., Berent, R., & Packman, D. (1982). The use of mathematical structure by inner city children. *Focus on Learning Problems in Mathematics*, *4*(2), 5–13.

Baroody, A. J., & Gannon, K. (1984). The development of the commutativity principle and economical addition strategies. *Cognition and Instruction*, *1*, 321–339.

Baroody, A. J., Gannon, K. E., Berent, R., & Ginsburg, H. P. (1984). The development of basic formal math abilities. *Acta Paedologica*, *1*, 133–150.

Baroody, A. J., & Ginsburg, H. P. (1982a). Generating number combinations: Rote process or problem solving? *Problem Solving*, *4*(12), 3–4.

Baroody, A. J., & Ginsburg, H. P. (1982b). Preschoolers' informal mathematical skills: Research and diagnosis. *American Journal of Diseases of Children*, *136*, 195–197.

Baroody, A. J., & Ginsburg, H. P. (1983). The effects of instruction on children's concept of "equals." *Elementary School Journal*, *84*,199–212.

Baroody, A. J., & Ginsburg, H. P. (1984, April). *TMR and EMR children's ability to learn counting skills and principles*. Paper presented at the annual meeting of the American Educational Research Association, New Orleans.

Baroody, A. J., & Ginsburg, H. P. (in press). The relationship between initial meaningful and mechanical knowledge of arithmetic. In J. Hiebert (Ed.), *Conceptual and procedural knowledge: The case of mathematics*. Hillsdale, NJ: Lawrence Erlbaum Associates.

Baroody, A. J., Ginsburg, H. P., & Waxman, B. (1983). Children's use of mathematical structure. *Journal for Research in Mathematics Education*, *14*, 156–168.

Baroody, A. J., & Mason, C. A. (1984). The case of Brian: An additional explanation for production deficiencies. In J. Moser (Ed.), *Proceedings of the Sixth Annual Meeting of the North American Chapter of the International Group for the Psychology of Mathematics Education* (pp. 2–8). Madison: Wisconsin Center for Educational Research.

Baroody, A. J., & Price, J. (1983). The development of the number-word sequence in the counting of three-year-olds. *Journal of Research in Mathematics Education, 14*, 361–368.

Baroody, A. J., & Snyder, P. (1983). A cognitive analysis of basic arithmetic abilities of TMR children. *Education and Training of the Mentally Retarded, 18*, 253–259.

Baroody, A. J., & White, M. (1983). The development of counting skills and number conservation. *Child Study Journal*, 13, 95–105.

Barsch, R. H. (1965). A movigenic curriculum. Madison: Wisconsin State Dept. of Public Instruction, No. 25.

Bearison, D. (1969). Role of measurement operations in the acquisition of conservation. *Developmental Psychology, 1*, 653–660.

Bebout, H. C. (1986, April). *Children's symbolic representation of addition and subtraction verbal problems.* Paper presented at the annual meeting of the American Education Research Association, San Francisco.

Beckmann, H. (1924). Die Entwicklung der Zahlleistung bei 2–6-jährigen Kindern. *Zeitschrift für Angewandte Psychologie, 22*, 1–72.

Beckwith, M., & Restle, F. (1966). Process of enumeration. *Psychological Review, 73*, 437–444.

Behr, M., Erlwanger, S., & Nichols, E. (1980). How children view the equals sign. *Mathematics Teaching, 92*, 13–15.

Behr, M. J., Lesh, R., Post, T. R., & Silver, E. A. (1983). Rational-number concepts. In R. Lesh & M. Landau (Eds.), *Acquisition of mathematics concepts and processes* (pp. 91–126). New York: Academic Press.

Bell, M. S. (1974). What does "everyman" really need from school mathematics? *Mathematics Teacher, 67*, 196–202.

Berent, R. (1982). *A critical analysis of the KeyMath Diagnostic Arithmetic Test.* Unpublished master's thesis, University of Rochester, Rochester, NY.

Binet, A. (1969). The perception of lengths and numbers. In R. H. Pollack and M. W. Brenner (Eds.), *The experimental psychology of Alfred Binet* (pp. 79–92). New York: Springer-Verlag.

Bisanz, J., Lefevre, J., Scott, C., & Champion, M. A. (1984, April). *Developmental changes in the use of heuristics in simple arithmetic problems.* Paper presented at the annual meeting of the American Educational Research Association, New Orleans.

Bjonerud, C. E. (1960). Arithmetic concepts possessed by the preschool child. *Arithmetic Teacher, 7*, 347–350.

Bley, N. S., & Thornton, C. A. (1981). *Teaching mathematics to the learning disabled.* Rockville, MD: Aspen.

Brainerd, C. J. (1973, March). The origins of number concepts. *Scientific American*, 101–109.

Bright, G., Harvey, J., & Wheeler, M. M. (1985). *Learning and mathematics games.* (*JRME* Monograph No. 1). Reston, VA: National Council of Teachers of Mathematics.

Brown, J. S., & Burton, R. R. (1978). Diagnostic models for procedural bugs in basic mathematical skills. *Cognitive Science*, *2*, 155–192.

Browne, C. E. (1906). The psychology of the simple arithmetical processes: A study of certain habits of attention and association. *American Journal of Psychology*, *17*, 2–37.

Brownell, W. A. (1935). Psychological considerations in the learning and the teaching of arithmetic. In D. W. Reeve (Ed.), *The teaching of arithmetic* (Tenth Yearbook, National Council of Teachers of Mathematics, pp. 1–31). New York: Bureau of Publications, Teachers College, Columbia University.

Brownell, W. A., & Chazal, C. (1935). The effects of premature drill in third-grade arithmetic. *Journal of Educational Research*, *29*, 17–28.

Bruner, J. S. (1963). *The process of education*. Cambridge: Harvard University Press.

Bruner. J. S. (1966). *Toward a theory of instruction*. Cambridge: Belknap Press of Harvard University Press.

Brush, L. (1978). Preschool children's knowledge of addition and subtraction. *Journal for Research in Mathematics Education*, *9*, 44–54.

Bunt, L. N. H., Jones, P. S., & Bedient, J. D. (1976). *The historical roots of elementary mathematics*. Englewood Cliffs, NJ: Prentice-Hall.

Buswell, G. T., & Judd, C. H. (1925). Summary of educational investigations relating to arithmetic. *Supplementary Educational Monographs*, No. 27. Chicago: University of Chicago Press.

Byers, V., & Herscovics, N. (1977). Understanding school mathematics. *Mathematics Teaching*, *81*, 24–24.

Campbell, J. I. D., & Graham, D. J. (1985). Mental multiplication skill: Structure, process, and acquisition. *Canadian Journal of Psychology*, *39*, 338–362.

Carpenter, T. P. (1985). Research on the role of structure in thinking. *Arithmetic Teacher*, *32*(6), 58–59.

Carpenter, T. P. (in press). Conceptual knowledge as a foundation for procedural knowledge: Implications from research on the initial learning of arithmetic. In J. Hiebert (Ed.), *Conceptual and procedural knowledge: The case of mathematics*. Hillsdale, NJ: Lawrence Erlbaum Associates.

Carpenter, T. P., & Bebout, H. C. (1985, April). *The representation of basic addition and subtraction word problems*. Paper presented at the annual meeting of the American Educational Research Association, Chicago.

Carpenter, T. P., Coburn, T. G., Reys, R. E., & Wilson, J. W. (1976). Notes from national assessment: Estimation. *Arithmetic Teacher*, *23*(4), 297–302.

Carpenter, T. P., Corbitt, M. K., Kepner, H. S., Lindquist, M. M., & Reys, R. E. (1980). Solving verbal problems: Results and implications from National Assessment. *Arithmetic Teacher*, *28*(1), 8–12.

Carpenter, T. P., Hiebert, J., & Moser, J. M. (1981). Problem structure and first grade children's initial solution processes for simple addition and subtraction problems. *Journal for Research in Mathematics Education*, *12*, 27–39.

Carpenter, T. P., Hiebert, J., & Moser, J. M. (1983). The effects of instruction on children's solutions of addition and subtraction word problems. *Educational Studies in Mathematics*, *14*, 55–72.

Carpenter, T. P., Lindquist, M. M., Matthews, W., & Silver, E. A. (1983). Results of the Third NAEP Mathematics Assessment: Secondary school. *Mathematics Teacher*, *76*, 652–660.

Carpenter, T. P., Matthews, W., Lindquist, M. M., & Silver, E. A. (1984). Achievement in mathematics: Results from the National Assessment. *Elementary School Journal*, *84*, 485–495.

Carpenter, T. P., & Moser, J. M. (1982). The development of addition and subtraction problem-solving skills. In T. P. Carpenter, J. M. Moser, & T. A. Romberg (Eds.), *Addition and subtraction: A cognitive perspective* (pp. 9–24). Hillsdale, NJ: Lawrence Erlbaum Associates.

Carpenter, T. P., & Moser, J. M. (1983). The acquisition of addition and subtraction concepts. In R. Lesh & M. Landau (Eds.), *Acquisition of mathematical concepts and processes* (pp. 7–44). New York: Academic Press.

Carpenter, T. P., & Moser, J. M. (1984). The acquisition of addition and subtraction concepts in grades one through three. *Journal for Research in Mathematics Education*, *15*, 179–202.

Carpenter, T. P. (in press). Conceptual knowledge as a foundation for procedural knowledge: Implications from research on the initial learning of arithmetic. In J. Hiebert (Ed.), *Conceptual and procedural knowledge: The case of mathematics*. Hillsdale, NJ: Lawrence Erlbaum Associates.

Carrison, D., & Werner, H. (1943). Principles and methods of teaching arithmetic to mentally retarded children. *American Journal of Mental Deficiency*, *47*, 309–317.

Carter, H. L. (1986). Linking estimation to psychological variables in the early years. In H. L. Schoen & M. J. Zweng (Eds.), *Estimation and mental computation* (pp. 74–81). Reston, VA: National Council of Teachers of Mathematics.

Churchill, E. M. (1961). *Counting and measuring: An approach to number education in the infant school*. Toronto: University of Toronto Press.

Cobb, P. (1985a). A reaction to three early number papers. *Journal for Research in Mathematics Education*, *16*, 141–145.

Cobb, P. (1985b). Two children's anticipations, beliefs, and motivations. *Educational Studies in Mathematics*, *16*, 111–126.

Connolly, A. J., Nachtman, W., & Pritchett, E. M. (1971/1976). *KeyMath diagnostic arithmetic test manual*. Circle Pines, MN: American Guidance Service.

Copeland, R. W. (1979). *How children learn mathematics*. New York: Macmillan.

Court, S. R. A. (1920). Numbers, time, and space in the first five years of a child's life. *Pedagogical Seminary*, *27*, 71–89.

Cruickshank, W. M. (1948). Arithmetic work habits of mentally retarded boys. *American Journal of Mental Deficiency*, *52*, 318–330.

Dantzig, T. (1930/1954). *Number: The language of science*. New York: The Free Press.

Dasen, P. R. (1972). Cross-cultural Piagetian research: A summary. *Journal of Cross-Cultural Psychology 3*, 23–39.

Davis, E. J., & McKillip, W. D. (1980). Improving story-problem solving in elementary school mathematics. In S. Krulik & R. E. Reys (Eds.), *Problem solving in school mathematics* (pp. 80–91). Reston, VA: National Council of Teachers of Mathematics.

Davis, P. J., & Hersh, R. (1981). *The mathematical experience*. Boston: Houghton Mifflin.

Davis, R. B. (in press). Conceptual and procedural knowledge in mathematics: A summary analysis. In J. Hiebert (Ed.), *Conceptual and procedural knowledge: The case of mathematics*. Hillsdale, NJ: Lawrence Erlbaum Associates.

Delacato, C. H. (1966). *Neurological organization and reading*. Springfield, IL: Charles C. Thomas.

de Lemos, M. (1969). The development of conservation in Aboriginal children. *International Journal of Psychology, 4*, 255–269.

Descoeudres, A. (1928). *The education of mentally defective children*. Boston: Heath.

Dewey, J. (1898). Some remarks on the psychology of number. *Pedagogical Seminary, 5*, 416–434.

Dewey, J. (1963). *Experience and education*. New York: Collier.

Dodwell, P. (1960). Children's understanding of number and related concepts. *Canadian Journal of Psychology, 14*, 191–205.

Dodwell, P. (1962). Relationship between the understanding of the logic of classes and of cardinal number in children. *Canadian Journal of Psychology, 16*, 152–160.

Driscoll, M. J. (1981). Research within reach: Elementary school mathematics. Reston, VA: National Council of Teachers of Mathematics.

Duckworth, E. (1982). A case study about some depths and perplexities of elementary arithmetic. In J. Bamberger & E. Duckworth (Eds.), *An analysis of data from an experiment in teacher development* (Grant No. G81-0042, pp. 44–170). Washington, DC: National Institute of Education.

Dumas, E., & Schminke, C. W. (1977). *Math activities for child involvement*. Boston: Allyn and Bacon.

Easley, J. A. (1983). Japanese approach to arithmetic. *For the Learning of Mathematics, 3*(3), 8–14.

Eccles, J., Adler, J. F., Futterman, R., Goff, S. B., Kaczala, C. M., Meece, J. L., & Midgley, C. (1983). Expectancies, values, and academic behaviors. In J. Spence (Ed.), *Achievement and achievement motivation* (pp. 75–146). San Francisco: W. H. Freeman.

Elkind, D. (1964). Discrimination, seriation, and numeration of size and dimensional difference in young children: Piagetian replication study VI. *Journal of Genetic Psychology, 104*, 275–296.

Ellis, A., & Harper, R. A. (1975). *A new guide to rational living*. North Hollywood, CA: Wilshire.

Engelhardt, J. M., Ashlock, R. B., & Wiebe, J. H. (1984). *Helping children understand and use numerals*. Boston: Allyn and Bacon.

Erlwanger, M. (1973). Benny's concept of rules and answers in IPI mathematics. *Journal of Children's Mathematical Behavior, 1*, 7–26.

Fennema, E., & Peterson, P. L. (1983, April). *Autonomous learning behavior: A possible explanation*. Paper presented at the annual meeting of the American Educational Research Association, Montreal.

Fernald, G. M. (1943). *Remedial techniques in basic school subjects*. New York: McGraw-Hill.

Fey, J. T. (1979). Mathematics teaching today: Perspectives from three national surveys. *Arithmetic Teacher*, 27(2), 10–14.

Flavell, J. H. (1970). Developmental studies of mediated memory. In H. W. Reese & L. P. Lipsitt (Eds.), *Advances in child development and behavior* (Vol. 5, pp. 181–211). New York: Academic Press.

Folsom, M. (1975). Operations on whole number. In J. N. Payne (Ed.), *Mathematics learning in early childhood* (pp. 162–190). Reston, VA: National Council of Teachers of Mathematics.

Frostig, M., LeFever, D. W., & Whittlesey, J. R. (1964). *The Marianne Frostig developmental test of visual perception*. Palo Alto, CA: Consulting Psychologists Press.

Furth, H. G., & Wachs, H. (1974). *Thinking goes to school: Piaget's theory in practice*. New York: Oxford University Press.

Fuson, K. C. (1982). An analysis of the counting-on solution procedure in addition. In T. P. Carpenter, J. M. Moser, & T. A. Romberg (Eds.), *Addition and subtraction: A cognitive perspective* (pp. 67-82). Hillsdale, NY: Lawrence Erlbaum Associates.

Fuson, K. C. (1984). More complexities in subtraction. *Journal of Research in Mathematics Education*, *15*, 214–225.

Fuson, K. C. (1985, March). *Teaching an efficient method of addition*. Paper presented at the annual meeting of the American Educational Research Association, Chicago.

Fuson, K. C. (in press-a). *Children's counting and concepts of number* New York: Springer-Verlag.

Fuson, K. C. (in press-b). Teaching children to subtract by counting up. *Journal for Research in Mathematics Education*.

Fuson, K. C., & Hall, J. W. (1983). The acquisition of early number word meanings: A conceptual analysis and review. In H. P. Ginsburg (Ed.), *The development of mathematical thinking* (pp. 49–107). New York: Academic Press.

Fuson, K. C., & Mierkiewicz, D. B. (1980, April). *A detailed analysis of the act of counting*. Paper presented at the annual meeting of the American Educational Research Association, Boston.

Fuson, K. C., Pergament, G. G., Lyons, B. G., & Hall, J. W. (1985). Children's conformity to the cardinality rule as a function of set size and counting accuracy. *Child Development*, *56*, 1429–1436.

Fuson, K. C., Richards, J., & Briars, D. J. (1982). The acquisition and elaboration of the number word sequence. In C. Brainerd (Ed.), *Children's logical and mathematical cognition: Progress in cognitive development* (pp. 33–92). New York: Springer-Verlag.

Gelman, R. (1972). The nature and development of early number concepts. In H. W. Reese (Ed.), *Advances in child development and behavior* (Vol. 7, pp. 115–167). New York: Academic Press.

Gelman, R. (1977). How young children reason about small numbers. In N. J. Castellan, D. B. Pisoni, & G. R. Potts (Eds.), *Cognitive Theory* (Vol. 2, pp. 219–238). Hillsdale, NJ: Lawrence Erlbaum Associates.

Gelman, R. (1982). Basic numerical abilities. In R. J. Sternberg (Ed.), *Advances in the psychology of intelligence* (Vol. 1, pp. 181–205). Hillsdale, NJ: Lawrence Erlbaum Associates.

Gelman, R., & Gallistel, C. R. (1978). *The child's understanding of number.* Cambridge, MA: Harvard University Press.

Gelman, R., & Meck, E. (1983). Preschoolers' counting: Principles before skill. *Cognition, 13*, 343–359.

Gelman, R., & Meck, E. (in press). The notion of principle: The case of counting. In J. Hiebert (Ed.), *Conceptual and procedural knowledge: The case of mathematics.* Hillsdale, NJ: Lawrence Erlbaum Associates.

Gibb, E. G. (1956). Children's thinking in the process of subtraction. *Journal of Experimental Education, 25*, 71–80.

Gibb, E. G., & Castaneda, A. M. (1975). Experience for young children. In J. N. Payne (Ed.), *Mathematics learning in early childhood* (pp. 96–124). Reston, VA: National Council of Teachers of Mathematics.

Gibson, E. J., & Levin, H. (1975). *The psychology of reading.* Cambridge: The M.I.T. Press.

Ginsburg, H. P. (1982). *Children's arithmetic.* Austin, TX: Pro-Ed.

Ginsburg, H. P., & Baroody, A. J. (1983). *The test of early mathematics ability.* Austin, TX: Pro-Ed.

Ginsburg, H. P., & Mathews, S. C. (1984). *Diagnostic test of arithmetic strategies.* Austin, TX: Pro-Ed.

Ginsburg, H. P., Posner, J. K., & Russell, R. L. (1981). The development of mental addition as a function of schooling. *Journal of Cross-Cultural Psychology, 12*, 163–178.

Ginsburg, H. P., & Russell, R. L. (1981). Social class and racial influences on early mathematical thinking. *Monographs of the Society for Research in Child Development, 46*, 16 (Serial No. 193).

Glaser, R. (1981). The future of testing: A research agenda for cognitive psychology and psychometrics. *American Psychologist, 36*, 923–936.

Gonchar, A. (1975). *A study in the nature and development of the natural number concept: Initial and supplementary analysis* (Technical Report No. 340). Wisconsin Research and Development Center for Cognitive Learning, University of Wisconsin, Madison, WI.

Goodnow, J., & Levine, R. A. (1973). "The grammar of action": Sequence and syntax in children's copying. *Cognitive Psychology, 4*, 82–98.

Greco, P., Grize, J., Papert, S., & Piaget, J. (1960). Problèmes de la construction du nombre. *Etudes d'épistémologie génétique.* Vol. 11. Paris: Presses Univer. France.

Green, R., & Laxon, V. (1970). The conservation of number, mother, water and a fried egg chez l'enfant. *Acta Psychologica, 32*, 1–20.

Groen, G., & Kieran, C. (1982). In search of Piagetian mathematics. In H. P. Ginsburg (Ed.), *The development of mathematical thinking* (pp. 351–375).

New York: Academic Press.
Groen, G. J., & Parkman, J. M. (1972). A chronometric analysis of simple addition. *Psychological Review*, *79*, 329–343.
Groen, G. J., & Resnick, L. B. (1977). Can preschool children invent addition algorithms? *Journal of Educational Psychology*, *69*, 645–652.
Hallahan, D., & Cruickshank, W. (1973). *Psychoeducational foundations of learning disabilities*. Englewood Cliffs, NJ: Prentice-Hall.
Hebbeler, K. (1977). Young children's addition. *Journal of Children's Mathematical Behavior*, *1*, 108–121.
Hiebert, J. (1984). Children's mathematics learning: The struggle to link form and understanding. *Elementary School Journal*, *84*, 497–513.
Hiebert, J., & Wearne, D. (1984, April). *A model of students' decimal computation procedures*. Paper presented at the annual meeting of the National Council of Teachers of Mathematics, San Francisco.
Holt, J. (1964). *How children fail*. New York: Delta.
Hood, H. B. (1962). An experimental study of Piaget's theory of the development of number in children. *British Journal of Psychology*, *53*, 273–286.
Hope, J. A. (1986). Mental calculation: Anachronism or basic skill? In H. L. Schoen & M. J. Zweng (Eds.), *Estimation and mental computation* (pp. 45–54). Reston, VA: National Council of Teachers of Mathematics.
House, P. A. (1980). Risking the journey into problem solving. In S. Krulik & R. E. Reys (Eds.), *Problem solving in school mathematics* (pp. 157–168). Reston, VA: National Council of Teachers of Mathematics.
Ilg, F., & Ames, L. B. (1951). Developmental trends in arithmetic. *The Journal of Genetic Psychology*, *79*, 3–28.
Jacobs, H. R. (1970). *Mathematics: A human endeavor*. San Francisco: W. H. Freeman.
James, W. (1939). *Talks to teachers on psychology*. New York: Henry Holt & Co.
Jerman, M. (1970). Some strategies for solving simple multiplication combinations. *Journal for Research in Mathematics Education*, *1*, 95–128.
Kamii, M. (1981). Children's ideas about written number. *Topics in Learning and Learning Disabilities*, *1*(3), 47–59.
Katona, G. (1940/1967). *Organizing and memorizing: Studies in the psychology of learning and teaching*. New York: Hafner.
Kiernan, C. (1980). The interpretation of the equal sign: Symbol for equivalence relations vs. an operator symbol. In R. Karplus (Ed.), *Proceedings of the Fourth International Conference for the Psychology of Mathematics Education* (pp. 163–169). Berkeley, CA: International Group for the Psychology of Mathematics Education.
Kilpatrick, J. (1985). Doing mathematics without understanding it: A commentary on Higbee and Kunihira. *Educational Psychologist*, *20*, 65–68.
Kirk, U. (1981). Learning to copy letters: A cognitive rule-governed task. *Elementary School Journal*, *81*, 29–33.
Klahr, D., & Wallace, J. G. (1973). The role of quantification operators in the development of conservation of quantity. *Cognitive Psychology*, *4*, 301–327.
Kline, M. (1974). *Why Johnny can't add*. New York: Vintage.

Kouba, V. (1986, April). *How young children solve multiplication and division word problems*. Paper presented at the National Council of Teachers of Mathematics research presession, Washington, DC.

Kraner, R. E. (1980). Math deficits of learning disabled first graders with mathematics as a primary and secondary disorder. *Focus on Learning Problems in Mathematics, 2*(3), 7–27.

Kratochwill, T. R., & Demuth, D. M. (1976). An examination of the predictive validity of the KeyMath diagnostic arithmetic test and the wide range achievement test in exceptional children. *Psychology in the Schools, 13*, 404–406.

Kulm, G. (1985). *Learning to add and subtract: Learning activities and implications from recent cognitive research* (National Institute of Education). Washington, DC: U. S. Government Printing Office.

LaPointe, K., & O'Donnell, J. (1974). Number conservation in children below age six: Its relationship to age, perceptual dimensions, and language comprehension. *Developmental Psychology, 10*, 422–428.

Lawson, G., Baron, J., & Siegel, L. (1974). The role of number and length cues in children's quantitative judgments. *Child Development, 45*, 731–736.

LeBlanc, J. F., Proudfit, L., & Putt, I. J. (1980). Teaching problem solving in the elementary school. In S. Krulik & R. E. Reys (Eds.), *Problem solving in school mathematics* (pp. 104–116). Reston, VA: National Council of Teachers of Mathematics.

Lerch, H. H. (1981). *Active learning experiences for teaching elementary school mathematics*. Boston: Houghton-Mifflin.

Lester, F. K., Jr. (1980). Research on mathematical problem solving. In R. J. Shumway (Ed.), *Research in mathematics education* (pp. 286–323). Reston, VA: National Council of Teachers of Mathematics.

Lester, F. K. (1983). Trends and issues in mathematical problem solving research. In R. Lesh & M. Landau (Eds.), *Acquisition of mathematics concepts and processes* (pp. 229–261). New York: Academic Press.

Leutzinger, L. P., Rathmell, E. C., & Urbatsch, T. D. (1986). Developing estimation skills in the primary grades. In H. L. Schoen & M. J. Zweng (Eds.), *Estimation and mental computation* (pp. 82-92). Reston, VA: National Council of Teachers of Mathematics.

Lindquist, M. M. (1984). The elementary school mathematics curriculum: Issues for today. *Elementary School Journal, 84*, 595–608.

Lindvall, C. M., & Ibarra, C. G. (1979, April). *The relationship of mode of presentation and of school/community differences to the ability of kindergarten children to comprehend simple story problems*. Paper presented at the annual meeting of the American Educational Research Association, Boston.

Lindvall, C. M., & Ibarra, C. G. (1980, April). *A clinical investigation of the difficulties evidenced by kindergarten children in developing "models" for the solution of arithmetic story problems*. Paper presented at the annual meeting of the American Educational Research Association, Boston. (ERIC Document Reproduction Service No. ED 193077)

Lunkenbein, D. (1985, April). *Cognitive structures underlying processes and conceptions in geometry*. Paper presented at the research presession of the annual meeting of the National Council of Teachers of Mathematics, San Antonio, TX.

Macnamara, J. (1975). A note on Piaget and number. *Child Development, 46*, 424–429.

Maffel, A. C., & Buckley, P. (1980). *Teaching preschool math: Foundations and activities*. New York: Human Sciences Press.

McCloskey, M., Caramazza, A., & Basili, A. (1984, February). *Dissociations of calculation processes*. Paper presented at the Institute of Naval Studies Symposium on Dyscalculia, Houston, TX.

McLuhan, M. (1964). *Understanding media: The extensions of man*. New York: McGraw-Hill.

Moyer, M. B., & Moyer, J. C. (1985). Ensuring that practice makes perfect: Implications for children with learning difficulties. *Arithmetic Teacher, 33*(1), 40–42.

Mpiangu, B., & Gentile, J. R. (1970). Is conservation of number a necessary condition for mathematical understanding? *Journal for Research in Mathematics Education, 1*, 179–192.

National Assessment of Educational Progress. (1983). *The third national mathematics assessment: Results, trends, and issues*. Denver: Education Commission of the States.

National Council of Supervisors of Mathematics. (1977). Position paper on basic mathematical skills. Washington, DC: National Institute of Education. See also (1977) *Arithmetic Teacher, 25*, pp. 19–22.

National Council of Teachers of Mathematics. (1980). *An agenda for action: Recommendations for school mathematics of the 1980's*. Reston, VA: National Council of Teachers of Mathematics.

Nichols, E. D., Anderson, P. A., Dwight, L. A., Flournoy, F., Kalin, R., Schluep, J., & Simon, L. (1978). *Holt school mathematics, book 8*. New York: Holt, Rinehart & Winston.

Noddings, N. (1985, April). *How formal should school mathematics be?* Invited address presented at the annual meeting of the American Educational Research Association, Chicago.

Olander, H. T. (1931). Transfer of learning in simple addition and subtraction. 2. *Elementary School Journal, 31*, 427–437.

Orton, S. T. (1937). *Reading, writing, and speech problems in children*. London: Chapman & Hall.

Parkman, J. M., & Groen, G. J. (1971). Temporal aspects of simple addition and comparison. *Journal of Experimental Psychology, 89*, 335–342.

Payne, J. N., & Rathmell, E. C. (1975). Number and numeration. In J. N. Payne (Ed.), *Mathematics learning in early childhood* (pp. 125–160). Reston, VA: National Council of Teachers of Mathematics.

Piaget, J. (1964). Development and learning. In R. E. Ripple & V. N. Rockcastle (Eds.), *Piaget rediscovered* (pp. 7–20). Ithaca, NY: Cornell University.

Piaget, J. (1965). *The child's conception of number*. New York: Norton.

Piaget, J. (1977). The role of action in the development of thinking. In W. F. Overton & J. M. Gallagher (Eds.), *Knowledge and development* (Vol. 1, pp.

17–42). New York: Plenum.

Polya, G. (1973). *How to solve it* (39th ed.). Princeton, NJ: Princeton University Press.

Quintero, A. H. (1985). Conceptual understanding of multiplication: Problems involving combination. *Arithmetic Teacher*, *33*(3), 36–39.

Rathmell, E. C. (1978). Using thinking strategies to teach basic facts. In M. N. Suydam & R. E. Reys (Eds.), *Developing computational skills* (pp. 13–50). Reston, VA: National Council of Teachers of Mathematics.

Resnick, L. B. (1982). Syntax and semantics in learning to subtract. In T. P. Carpenter, J. M. Moser, & T. A. Romberg (Eds.), *Addition and subtraction: A cognitive perspective* (pp. 136–155). Hillsdale, NJ: Lawrence Erlbaum Associates.

Resnick, L. B. (1983). A developmental theory of number understanding. In H. P. Ginsburg (Ed.), *The development of mathematical thinking* (pp. 109–151). New York: Academic Press.

Resnick, L. B., & Ford, W. W. (1981). *The psychology of mathematics for instruction*. Hillsdale, NJ: Lawrence Erlbaum Associates.

Resnick, L. B., & Neches, R. (1984). Factors affecting individual differences in learning ability. In R. J. Sternberg (Ed.), *Advances in the psychology of human intelligence* (Vol. 2, pp. 275–323). Hillsdale, NJ: Lawrence Erlbaum Associates.

Reyes, L. H. (1984). Affective variables and mathematics education. *Elementary School Journal*, *84*, 558–581.

Reys, B. J. (1986). Teaching computational estimation: Concepts and strategies. In H. L. Schoen & M. J. Zweng (Eds.), *Estimation and mental computation* (pp. 16–30). Reston, VA: National Council of Teachers of Mathematics.

Reys, R. E. (1984). Mental computation and estimation: Past, present, and future. *Elementary School Journal*, *84*, 544–557.

Riley, M. S., Greeno, J. G., & Heller, J. I. (1983). Development of children's problem-solving ability in arithmetic. In H. P. Ginsburg (Ed.), *The development of mathematical thinking* (pp. 153–200). New York: Academic Press.

Romberg, T. A. (1982). An emerging paradigm for research on addition and subtraction skills. In T. P. Carpenter, J. M. Moser, & T. A. Romberg (Eds.), *Addition and subtraction: A cognitive perspective* (pp. 1–7). Hillsdale, NJ: Lawrence Erlbaum Associates.

Romberg, T. A. (1984, April). *School mathematics: Options for the 1990s. Chairman's Report of a Conference* (U. S. Department of Education, Office of Educational Research and Improvement, National Council of Teachers of Mathematics, Wisconsin Center for Educational Research). Madison, WI (December 5–8, 1983).

Rosner, J. (1971a). *The design board program*. Pittsburgh: Learning Research and Development Center, University of Pittsburgh.

Rosner, J. (1971b). *Phonic analysis training and beginning reading skills*. Pittsburgh: Learning Research and Developmental Center, University of Pittsburgh.

Russell, B. (1917). *Introduction to mathematical philosophy*. London: George,

Allen, and Unwin.
Russell, R. L., & Ginsburg, H. P. (1984). Cognitive analysis of children's mathematics difficulties. *Cognition and Instruction*, *1*, 217–244.
Schaeffer, B., Eggleston, V., & Scott, J. (1974). Number development in young children. *Cognitive Psychology*, *6*, 357–379.
Schoenfeld, A. H. (1982). Some thoughts on problem-solving research and mathematics education. In F. K. Lester, Jr., & J. Garofalo (Eds.), *Mathematical problem solving: Issues in research* (pp. 27–37). Philadelphia: Franklin Institute Press.
Schoenfeld, A. H. (1985). *Mathematical problem solving*. New York: Academic Press.
School Mathematics Study Group. (1965). *Mathematics for the elementary school: Teacher's commentary*. New Haven: Yale University Press.
Secada, W. G., Fuson, K. C., & Hall, J. (1983). The transition from counting-all to counting-on in addition. *Journal for Research in Mathematics Education*, *14*, 47–57.
Sharp, E. (1969). *Thinking is child's play*. New York: E. P. Dutton.
Siegler, R. S., & Robinson, M. (1982). The development of numerical understandings. In H. W. Reese & L. P. Lipsitt (Eds.), *Advances in child development and behavior* (Vol. 1, pp. 241–312). New York: Academic Press.
Siegler, R. S., & Shrager, J. (1984). Strategy choices in addition: How do children know what to do? In C. Sophian (Ed.), *Origins of cognitive skills* (pp. 229–293). Hillsdale, NJ: Lawrence Erlbaum Associates.
Silver, E. A., & Thompson, A. G. (1984). Research perspective on problem solving in elementary school mathematics. *Elementary School Journal*, *84*, 529–545.
Sinclair, H., & Sinclair, A. (in press). Children's mastery of written numerals and the construction of basic number concepts. In J. Hiebert (Ed.), *Conceptual and procedural knowledge: The case of mathematics*. Hillsdale, NJ: Lawrence Erlbaum Associates.
Smith, D. E. (1923). *The history of mathematics*. Vol. 1. Boston, MA: Ginn.
Smith, J. H. (1921). Arithmetical combinations. *Elementary School Journal*, *10*, 762–770.
Spradlin, J. E., Cotter, V. M., Stevens, C., & Friedman, M. (1974). Performance of mentally retarded children on pre-arithmetic tasks. *American Journal of Mental Deficiency*, *78*, 397–403.
Starkey, P., & Cooper, R. (1977). *The role of estimation skills in the development of number conservation*. Paper presented at the Seventh Annual Symposium of the Jean Piaget Society, Philadelphia, PA.
Starkey, P., & Cooper, R. G. (1980). Perception of numbers by human infants. *Science*, *210*, 1033–1035.
Starkey, P., & Gelman, R. (1982). The development of addition and subtraction abilities prior to formal schooling in arithmetic. In T. P. Carpenter, J. M. Moser, & T. A. Romberg (Eds.)., *Addition and subtraction: A cognitive perspective* (pp. 99–116). Hillsdale, NJ: Lawrence Erlbaum Associates.
Starkey, P. D., Spelke, E. S., & Gelman, R. (in press). Numerical abstraction by human infants. *Cognition*.

Steffe, L. P., von Glasersfeld, E., Richards, J., & Cobb, P. (1983). *Children's counting types*. New York: Praeger.
Steinberg, R. M. (1985). Instruction on derived fact strategies in addition and subtraction. *Journal for Research in Mathmatics Education*, *16*, 337–355.
Strauss, A. A., & Lehtinen, L. E. (1950). *Psychopathology and education of the brain-injured child*. New York: Grune & Stratton.
Suydam, M., & Weaver, J. F. (1975). Research on mathematics learning. In J. N. Payne (Ed.), *Mathematics learning in early childhood* (37th Yearbook of the National Council of Teachers of Mathematics, pp. 43-67). Reston, VA: National Council of Teachers of Mathematics.
Svenson, O. (1975). Analyses of time required by children for simple additions. *Acta Psychologica*, *35*, 289–302.
Swenson, E. J. (1949). Organization and generalization as factors in learning, transfer, and retroactive ihibition. *Learning theory in school situations* (University of Minnesota Studies in Education No. 2). Minneapolis: University of Minnesota Press.
Thiele, C. (1938). *The contribution of generalization to the learning of the addition facts*. New York: Bureau of Publications, Teachers College, Columbia University.
Thorndike, E. L. (1922). *The psychology of arithmetic*. New York: Macmillan.
Thornton, C. A. (1978). Emphasizing thinking strategies in basic fact instruction. *Journal for Research in Mathematics Education*, *9*, 213–227.
Thornton, C. A., & Toohey, M. A. (1985). Basic math facts: Guidelines for teaching and learning. *Learning Disabilities Focus* (pp. 10–14).
Tobias, S. (1978). *Overcoming math anxiety*. New York: Norton.
Trafton, P. R. (1978). Estimation and mental arithmetic: Important components of computation. In M. N. Suydam & R. E. Reys (Eds.), *Developing computational skills* (pp. 196–213). Reston, VA: National Council of Teachers of Mathematics.
Trafton, P. R. (1986). Teaching computational estimation: Establishing an estimation mind-set. In H. L. Schoen & M. J. Zweng (Eds.), *Estimation and mental computation* (pp. 16–30). Reston, VA: National Council of Teachers of Mathematics.
Traub, N. (1977). *Recipe for reading*. New York: Walker.
Trivett, J. (1980). The multiplication table: To be memorized or mastered? *For the Learning of Mathematics*, *1*(1), 21–25.
Underhill, B., Uprichard, E., & Heddens, J. (1980). *Diagnosing mathematical difficulties*. Columbus: Charles E. Merrill.
Van de Walle, J. (1980). *An investigation of the concepts of equality and mathematical symbolism held by first, second, and third grade children: An informal report*. Paper presented at the national meeting of the National Council of Teachers of Mathematics, Seattle.
van Engen, H., & Grouws, D. (1975). Relations, number sentences, and other topics. In J. N. Payne (Ed.), *Mathematics learning in early childhood* (pp. 251–271). Reston, VA: National Council of Teachers of Mathematics.
Van Lehn, K. (1983). On the representation of procedures in repair theory. In H. P.

Ginsburg (Ed.), *The development of mathematical thinking* (pp. 197–252). New York: Academic Press.

Vellutino, F. R., Steger, B. M., Moyer, S. C., Harding, C. J., & Niles, J. A. (1977). Has the perceptual deficit hypothesis led us astray? *Journal of Learning Disabilities, 10*, 375–385.

von Glasersfeld, E. (1982). Subitizing: The role of figural patterns in the development of numerical concepts. *Archives de Psychologie, 50*, 191–218.

Wagner, S., & Walters, J. (1982). A longitudinal analysis of early number concepts: From numbers to number. In G. Formam (Ed.), *Action and thought* (pp. 137–161). New York: Academic Press.

Wang, M., Resnick, L., & Boozer, R. (1971). The sequence of development of some early mathematics behavior. *Child Development, 42*, 1767–1778.

Weaver, J. F. (1973). The symmetric property of the equality relation and young children's ability to solve open addition and subtraction sentences. *Journal for Research in Mathematics Education, 4*, 45–46.

Wertheimer, M. (1945). *Productive thinking*. New York: Harper & Row.

Wheeler, L. R. (1939). A comparative study of the difficulty of the 100 addition combinations. *The Journal of Genetic Psychology, 54*, 295–312.

Wohlwill, J., & Lowe, R. (1962). Experimental analysis of the development of the conservation of number. *Child Development, 33*, 153–167.

Woods, S. S., Resnick, L. B., & Groen, G. J. (1975). An experimental test of five process models for subtraction. *Journal of Educational Psychology, 67*, 17–21.

Wynroth, L. (1969/1980). *Wynroth math program – The natural numbers sequence*. Ithaca, NY: Wynroth Math Program.

Zimiles, H. (1963). A note on Piaget's concept of conservation. *Child Development, 34*, 691–695.

About the Author

Arthur J. Baroody was born and raised in Auburn, NY. He received a B.S. degree in science education in 1969 and a Ph.D. in educational psychology in 1979 from Cornell University. After serving stints with the U.S. Army, Liverpool (NY) Central Schools, George Junior Republic residential treatment center, and Keuka College, he went to the University of Rochester, where he was a Research Associate from 1980 to 1986. He is currently a faculty member of the Department of Elementary and Early Childhood Education, College of Education, University of Illinois at Urbana-Champaign. He married the former Sharon R. Coslick in 1970 and has three children.

Index